GW00633233

Progress in
Respiration Research

Vol. 15

Series Editor
H. Herzog, Basel

S. Karger · Basel · München · Paris · London · New York · Sydney

International Symposium on Clinical Importance of Surfactant Defects,
Hamburg, October 31 – November 2, 1979

Clinical Importance of Surfactant Defects

Volume Editor
P. von Wichert, Hamburg

153 figures and 38 tables, 1981

S. Karger · Basel · München · Paris · London · New York · Sydney

Progress in Respiration Research

Vol. 12: Cough and Other Respiratory Reflexes.
  Editors: *J. Korpàs* and *Z. Tomori,* Martin
  XII + 356 p., 102 fig., 2 tab., 1979. ISBN 3-8055-3007-2
Vol. 13: Pulmonary Embolism.
  Editor: *J. Widimský,* Prague
  VIII + 192 p., 78 fig., 36 tab., 1980. ISBN 3-8055-0487-X
Vol. 14: Asthma.
  Editor: *H. Herzog,* Basel
  X + 314 p., 100 fig., 33 tab., 1980. ISBN 3-8055-0991-X

National Library of Medicine, Cataloging in Publication
  International Symposium on Clinical Importance of Surfactant Defects, Hamburg
  1979. Clinical importance of surfactant defects / volume editor, P. von Wichert. –
  –Basel ; New York : Karger, 1981.
  (Progress in respiration research ; v. 15)
  1. Pulmonary Surfactant – physiology – congresses   2. Respiratory Distress
  Syndrome – etiology – congresses   I. Wichert, Peter von   ed.   II. Title   III. Series
  W3 PR948 v. 15 1979   WF 600 I6106c 1979
  ISBN 3-8055-1011-X

All rights reserved.
No part of this publication may be translated into other languages, reproduced or
utilized in any form or by any means, electronic or mechanical, including photo-
copying, recording, microcopying, or by any information storage and retrieval
system, without permission in writing from the publisher.

© Copyright 1981 by S. Karger AG, Basel (Switzerland)
  Printed in Switzerland by Werner Druck AG, Basel
  ISBN 3-8055-1011-X

Contents

Respiratory Distress Syndrome of the Newborn and the Adult

Therapeutical Approaches

Foreword

In recent years interest in the pulmonary surfactant system has very much increased and interesting experimental work has been done. Until now the theoretical view has dominated because the surfactant system plays a basic role in lung physiology. However, it has recently been acknowledged that disturbances in the surfactant system may be of great clinical importance. This was shown very impressively in hyaline membrane disease of the newborn, but there is an extending field of knowledge showing a pathogenetically important role of surfactant in adult diseases like embolism, lung edema and shock lung as well.

The present symposium, held in Hamburg on October 31st and November 1st 1979 discusses the clinical importance of the surfactant system. Different aspects of experimental work, scientific problems of anesthesia, intensive care, obstetrics and pediatrics have been touched upon. These different points of view have led to a mutual benefit of clinical and experimental work and many new ideas and insights into the function of the surfactant system and its role in the pathogenesis of lung diseases as well as possibilities in treating disorders of the surfactant system are presented and discussed.

The symposium was generously supported by: Senat der Freien und Hansestadt Hamburg; Deutsche Forschungsgemeinschaft; Jung-Stiftung für Wissenschaft und Forschung, Hamburg, and Dr. *Carl Thomae*, Biberach/Riss. This support also made it possible to publish the results of the symposium. The help of Prof. Dr. *H. Herzog*, Basel, editor of the series *Progress in Respiration Research* is very much appreciated.

Peter von Wichert

Basic Problems

Prog. Resp. Res., vol. 15, pp. 1–19 (Karger, Basel 1981)

Synthesis of Surfactant Lipids: Studies with Type II Alveolar Cells Isolated from Adult Rat Lung

J.J. Batenburg, M. Post and L.M.G. Van Golde

Laboratory of Veterinary Biochemistry, State University of Utrecht, Utrecht

Introduction

In 1972, *King and Clements* [39] reported in a classical paper that dipalmitoylphosphatidylcholine is the major active component of the pulmonary surfactant system which lines the alveoli and prevents their collapse during expiration. Evidence is accruing that other phospholipids such as phosphatidylglycerol are also essential components of pulmonary surfactant. In most tissues, phosphatidylglycerol occurs in minute amounts only, whereas this phospholipid represents about 10% of the surfactant lipids in the lung [28, 52]. Although the exact function of phosphatidylglycerol in surfactant has not yet been defined, its presence appears to be critical for preventing respiratory distress syndrome in the neonate [26]. Many investigators have performed studies on the pathways involved in the formation of pulmonary lipids, with particular emphasis on the biosynthesis of dipalmitoylphosphatidylcholine. Most of these investigations have, however, been carried out with whole lung or subcellular fractions derived thereof. Although a wealth of valuable information has come from such studies, they could not provide conclusive answers with respect to the detailed mechanisms operating in the synthesis of surfactant lipids and the regulation of these processes. The obvious reason for the relatively slow progress in this field was the enormous heterogeneity of lung tissue which consists of about 40 different cell types [60] whereas the synthesis of surfactant occurs solely in the alveolar type II cells [for reviews see 11, 24, 62]. It is only since the last few years that methods are available to isolate more or less homogeneous populations of these type II cells from normal lung [13, 14, 37, 38, 43–47].

After a short summary of our current knowledge with respect to the pathways involved in the biosynthesis of dipalmitoylphosphatidylcholine and phosphatidylglycerol, the present paper will focus on studies of these processes in isolated alveolar type II cells from adult rat lung as well as on the influence of corticosteroids on the formation of surfactant lipids in these cells. In addition, the isolated type II cell will be compared with the isolated perfused rat lung as model to study surfactant synthesis in the adult lung.

De novo *synthesis of Phosphatidylcholine and Phosphatidylglycerol*

Phosphatidic acid occupies an important branchpoint in the *de novo* synthesis of phosphatidylcholine and phosphatidylglycerol (fig. 1). It may be synthesized either via glycerol-3-phosphate or via acyl-dihydroxyace-tonephosphate (acyl-DHAP). Studies with subcellular fractions derived from whole lung [21] indicated that the route via acyl-DHAP is indeed a potential alternate to the glycerol-3-phosphate pathway. More recently, *Mason* [42] showed the presence of the acyl-DHAP pathway in isolated rat lung alveolar type II cells. Assuming the existence of one single glycerol-3-phosphate pool in type II cells, he calculated that 56% of the phosphatidylglycerol and 64% of the phosphatidylcholine was synthesized by means of the acyl-DHAP pathway.

Phosphatidate phosphohydrolase and phosphatidate cytidylyltransferase catalyze the conversion of phosphatidic acid into diacylglycerol and CDP-diacylglycerol, respectively (fig. 1). Diacylglycerol reacts with CDP-choline, which is synthesized from choline by the sequential action of choline kinase and cholinephosphate cytidylyltransferase, to yield phosphatidylcholine, a reaction which is catalyzed by cholinephosphotransferase. There is abundant evidence that this CDP-choline or *Kennedy* [35] pathway is the major route for the *de novo* synthesis of pulmonary phosphatidylcholine. Although N-methylation of phosphatidylethanolamine is demonstrable in the lung, it is now generally accepted that this pathway is of minor significance for the formation of phosphatidylcholine in the adult and fetal lung of both primates and nonprimates [for discussion see 11, 18, 62].

The formation of phosphatidylglycerol from CDP-diacylglycerol and glycerol-3-phosphate proceeds in two steps (fig. 1), as has been shown by *in vitro* studies with subcellular fractions from whole lung [27, 28]. It has

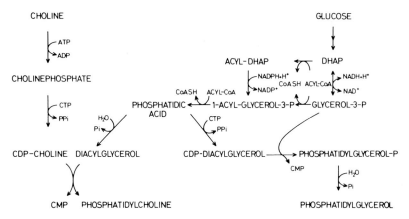

Fig. 1. *De novo* synthesis of phosphatidylcholine and phosphatidylglycerol.

been postulated that the conversion of phosphatidylglycerolphosphate into phosphatidylglycerol may be catalyzed by the same enzyme that is responsible for the formation of diacylglycerol from phosphatidic acid [33], although conclusive evidence for this postulate is still lacking.

Remodeling of Unsaturated Phosphatidylcholines into Dipalmitoyl-phosphatidylcholine

As has been extensively discussed in earlier reviews [11, 62], the CDP-choline pathway produces in the lung predominantly phosphatidylcholines containing palmitic acid at the 1-position and an unsaturated fatty acid at the 2-position. These *de novo* synthesized unsaturated phosphatidylcholines can be remodeled into dipalmitoylphosphatidylcholine by means of at least two mechanisms (fig. 2): (a) a deacylation-reacylation cycle and (b) a deacylation-transacylation process.

Both processes require as the first step removal of the unsaturated fatty acid by a phospholipase A_2. There have been several reports describing the presence of phospholipase A_2 activity in the soluble portion [49] and microsomes [23] of rat lung. *Longmore et al.* [41] reported recently that the microsomal phospholipase A_2 from rat lung displayed a pronounced preference to degrade membrane-bound phosphatidylcholines which contained an unsaturated fatty acid at the 2-position. Such preference

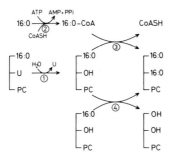

Fig. 2. Remodeling of *de novo* synthesized unsaturated phosphatidylcholine into dipalmitoylphosphatidylcholine. 'U' stands for an unsaturated fatty acyl group. Enzymes: 1 = phospholipase A_2; 2 = acyl-CoA synthetase; 3 = lysolecithin acyltransferase; 4 = lysolecithin: lysolecithin acyltransferase.

would be in line with a physiological role of this enzyme in the remodeling of unsaturated phosphatidylcholines. The 1-palmitoyl-*sn*-glycero-3-phosphocholine produced by the phospholipase A_2 can be reacylated with palmitoyl-CoA to yield dipalmitoylphosphatidylcholine, a reaction which is catalyzed by lysolecithin acyltransferase [40]. Lysolecithin:lysolecithin acyltransferase [16] catalyzes the transfer of the palmitoyl moiety of one molecule of 1-palmitoyl-*sn*-glycero-3-phosphocholine to a second 1-palmitoyl-*sn*-glycero-3-phosphocholine, which results in the formation of dipalmitoylphosphatidylcholine and glycero-3-phosphocholine.

Both reacylation [22, 63] and transacylation [1, 12, 29] proceed in preparations derived from whole lung with sufficient specificity to be responsible for the synthesis of dipalmitoylphosphatidylcholine, but the relative importance of both processes in the production of surfactant dipalmitoylphosphatidylcholine was not established by these studies. Recent investigations [8] with isolated alveolar type II cells from adult lung provided, however, evidence that reacylation is more important than transacylation for the synthesis of surfactant dipalmitoylphosphatidylcholine in the adult lung (see below).

Isolation of Type II Cells

Several methods have been developed for the isolation of type II alveolar epithelial cells from normal lung. After enzymatic dispersal of

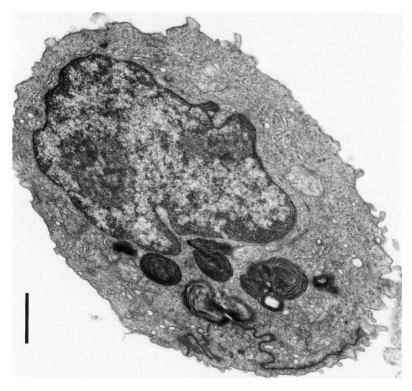

Fig. 3. Electron micrograph of a type II cell isolated from normal adult rat lung after 20 h in culture. Bar = 1 μm.

lung tissue type II cells can be obtained from the resultant heterogeneous cell suspension by centrifugation [37, 38, 47], cloning [13], or differential attachment to a solid support [14, 44–46]. In the present study type II cells were isolated from normal adult rats by trypsinization followed by density gradient centrifugation and differential adherence in primary monolayer culture as described in detail by *Mason et al.* [44–47]. Per preparation of 4 rats the yield was $8 \pm 2 \times 10^6$ cells. The percentage type II cells was routinely estimated by phosphine 3R [47] and the modified Papanicolaou stain [37] and appeared to be $96 \pm 2\%$ which was in good agreement with results from electron microscopy. The viability of the type II cells, as judged by trypan blue exclusion, was $95 \pm 3\%$. A transmission electron micrograph of a type II cell isolated in our laboratory is shown in figure 3.

Influence of Fatty Acids on the Synthesis of Phosphatidylcholine in Isolated Type II Cells

In view of their putative role as producers of surfactant it is a prerequisite that type II cells have the capacity to synthesize phosphatidylcholines with a high percentage of the disaturated species. *Batenburg et al.* [9] incubated rat type II cells, isolated as described above, in the presence of optimal concentrations of [1–14C]acetate (*plus* 5.6 mM unlabelled glucose), [1-14C]palmitate (*plus* 5.6 mM glucose), [*Me*–14C]choline (*plus* 5.6 mM glucose and 0.2 mM palmitate), [U-14C]glucose (*plus* 0.2 mM choline and 0.2 mM palmitate) and [1,3-3H]glycerol (*plus* 0.2 mM choline and 0.2 mM palmitate). In all cases 74% or more of the phosphatidylcholines synthesized from the labelled precursors consisted of disaturated species. These findings extended studies by *Mason et al.* [43, 44] on rat type II cells, by *Kikkawa et al.* [36] and *Smith and Kikkawa* [58] on rabbit type II cells, and by *Wykle et al.* [64] on urethan-induced pulmonary adenoma from the mouse, which are used as model for isolated type II cells [59].

It should be realized that the experiments of *Batenburg et al.* [9] were all performed in the presence of unlabelled exogenous palmitate, except the experiment with [1-14C]acetate as precursor. The source of fatty acids for the biosynthesis of surfactant lipids by the type II cell *in vivo* is not known: the cells may take up fatty acids transported to the lung via the blood as free fatty acids [31] or as lipoproteins [30]. Therefore, it was deemed of interest to investigate the distribution of [U–14C]glucose among the various classes of phosphatidylcholine in the presence and absence of various exogenous fatty acids. In the absence of exogenous fatty acids, 60% of the phosphatidylcholines synthesized by the type II cells from [U-14C]glucose, was disaturated (fig. 4). This percentage was enhanced by the addition of palmitate at the expense of the monoenoic and dienoic phosphatidylcholines. Addition of oleate and linoleate resulted in decreased percentages of labelled disaturated phosphatidylcholines and increased percentages of labelled mono- and dienoic phosphatidylcholines, respectively. In the simultaneous presence of exogenous palmitate and oleate, the percentage disaturated phosphatidylcholine synthesized was in between that measured in the presence of palmitate alone and that found with oleate alone. These results showed that the composition of the medium is reflected in the partition of glucose among the various molecular classes of phosphatidylcholine. This would suggest that *in vivo* the fatty acid

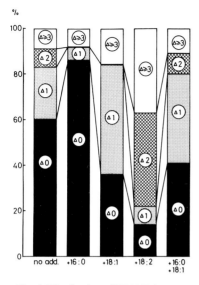

Fig. 4. Distribution of [U-¹⁴C]glucose among the molecular classes of phosphatidylcholine after incubation of adult rat lung type II cells in the presence and absence of various fatty acids [for details see 9].

concentration of the blood as related to the amount of glucose and the degree of unsaturation of the fatty acids are factors controlling the synthesis of surfactant lipids.

Remodeling of Unsaturated Phosphatidylcholines into Disaturated Phosphatidylcholines in Isolated Type II Cells

As shown in figure 2, auxiliary mechanisms such as deacylation-reacylation or deacylation-transacylation are required to transform *de novo* synthesized unsaturated phosphatidylcholines into dipalmitoylphosphatidylcholine. This information had been obtained from studies on whole lung [for review see 62]. Using adenomas from the mouse as a model *Wykle et al.* [64] provided evidence that remodeling of unsaturated phosphatidylcholines also occurs in type II cells. These authors suggested that the deacylation-reacylation cycle might be responsible for the remodeling in the type II cell, a suggestion which was supported by studies on the substrate specificity of lysolecithin acyltransferase in long-term cultures of rabbit-

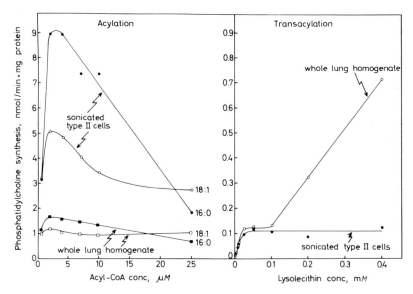

Fig. 5. Activities of lysolecithin acyltransferase and lysolecithin:lysolecithin acyltransferase in type II cell sonicate and in whole lung homogenate as a function of the acyl-CoA and lysolecithin concentration, respectively [for experimental details see 8].

lung cells presumably derived from type II cells [61]. In these studies [61, 64] no comparison was made between the activity of lysolecithin acyltransferase and that of lysolecithin:lysolecithin acyltransferase, the enzyme catalyzing the transacylation.

Figure 5 shows a comparison of the activities of lysolecithin acyltransferase and lysolecithin:lysolecithin acyltransferase in fresh type II cells from normal lung. The activity of lysolecithin acyltransferase in sonicated type II cells is far greater than that in whole lung homogenate, both with palmitoyl-CoA and with oleoyl-CoA as the acyl donor. Over a wide range of substrate concentrations, the enzyme displays a preference for palmitoyl-CoA as a substrate. If measured in the presence of 10 mM Mg^{2+} the activity towards palmitoyl-CoA is higher at all concentrations even up to 50 µM [8]. It is also possible to compare the incorporation of free palmitate and oleate. In this case the fatty acids have to be activated by acyl-CoA synthetase of the type II cell sonicate itself (fig. 2, reaction 2) before lysolecithin acyltransferase can catalyze their entry into phosphatidylcholine (fig. 2, reaction 3). Also under this assay condition, type II cell sonicate incorporates palmitate faster into phosphatidylcholine than oleate [data not shown, see 8].

Lysolecithin:lysolecithin acyltransferase shows normal saturation kinetics in type II cells, in contrast to the enzyme in whole lung. At concentrations of 0.1 mM or lower the specific activity of lysolecithin:lysolecithin acyltransferase in whole lung homogenate is about equal to that in type II cell sonicate. At concentrations exceeding 0.1 mM, the specific activity in type II cells is much lower than that in whole lung homogenate.

It is important to note that the specific activity of lysolecithin:lysolecithin acyltransferase in type II cells is at least one order of magnitude lower than that of lysolecithin acyltransferase. The observations described above provide strong indications that in type II cells from normal adult rat lung the deacylation-reacylation cycle is more important for the synthesis of dipalmitoylphosphatidylcholine than the deacylation-transacylation process. Some caution, however, seems appropriate. *Finkelstein and Mavis* [20] reported in a recent abstract that in lung cells isolated after trypsin treatment some enzymes involved in phospholipid metabolism are significantly damaged. Although damaging effects on lysolecithin acyltransferase or lysolecithin:lysolecithin acyltransferase were not reported, different sensitivities of these two enzymes to trypsin treatment would affect the interpretation of the results presented in this study. It should be stressed, however, that the experiments of *Finkelstein and Mavis* [20] were carried out in cells directly after trypsin treatment, whereas the present results were obtained with cells that had been in culture for 20 h after trypsinization. In this light it is relevant to refer to the finding of *Kasten* [34] that the intracellular damage to cardiac cells caused by trypsin was repaired within 15 h after removal of trypsin. In addition, the type II cells isolated as described in this paper still have the capacity to synthesize phosphatidylcholines with a high degree of saturation [9]. This would not be expected if the enzyme responsible for the conversion of unsaturated phosphatidylcholines into dipalmitoylphosphatidylcholine would be damaged.

Effects of Cortisol on the Synthesis of Phosphatidylcholine and Phosphatidylglycerol by Isolated Type II Cells from Adult Rat Lung

There is abundant evidence that during the later part of gestation the maturation of the fetal lung is influenced by circulating glucocorticoste-

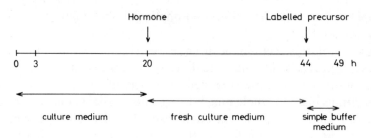

Fig. 6. Plating system for studying the effects of hormones on the synthesis of phospholipids in adult rat lung type II cells in primary culture. For details see text.

roids [cf. 5]. Administration of corticosteroids to the fetus accelerates lung
maturation and production of surfactant in a variety of mammalian
species [4, 17] and corticosteroids have also been used to prevent the onset
of respiratory distress syndrome in the human neonate [7].

The mechanisms via which corticosteroids accelerate the production
of surfactant in the developing lung are not yet fully understood. There is
evidence from studies on minces and slices of fetal rabbit lung [19, 53],
organ cultures of human [15] and rat [25] fetal lung, and monolayer
cultures of rabbit fetal lung cells [57] that corticosteroids promote the
synthesis of phosphatidylcholines from a variety of labelled precursors.
Less certainty exists on the question as to which of the enzymes involved
in the synthesis of phosphatidylcholines are affected by glucocorticosteroid
treatment [for discussion see 51]. Relatively few studies have focussed on
the effects of hormones on lipid metabolism in the adult lung, although
there is some evidence that corticosteroids also affect lung morphology and
lipid metabolism in the adult animal [2,50].

It is interesting that corticosteroid receptors have been demonstrated
in isolated type II cells from normal rat lung which indicates that type II
cells may be influenced by a direct action of these hormones [6]. Studies
with isolated type II cells have shown that corticosteroids promote the
formation of phosphatidylcholines from a variety of labelled precursors [3,
54, 55]. However, such studies have so far been confined to experiments
with cell-line A549, derived from a human lung carcinoma [54], and with
cells isolated from normal lung tissue by cloning [3, 55]. It is uncertain
whether these results may be directly extrapolated to normal type II cells
[for discussion see 11, 46]. In the present study it was decided to investigate
the effects of cortisol on the synthesis of phosphatidylcholines and phos-

Table I. Effects of cortisol on the incorporation of various labelled precursors into total phosphatidylcholines (PC), disaturated phosphatidylcholines (DSPC) and phosphatidyl-glycerols (PG) in adult rat lung type II cells

Lipid	Ratio cortisol-treated/control				
	[Me–14C]choline (n = 16)	[1–14C]acetate (n = 8)	[1–14C]palmitate (n = 10)	[U–14C]glucose (n = 10)	[1,3–3H]glycerol (n = 8)
PC	1.27 ± 0.05*	1.27 ± 0.04*	1.16 ± 0.03*	1.44 ± 0.10*	1.23 ± 0.05*
DSPC	1.29 ± 0.06*	1.27 ± 0.04*	1.22 ± 0.05*	1.51 ± 0.12*	1.37 ± 0.06*
PG		1.33 ± 0.05*	1.35 ± 0.03*	1.20 ± 0.04*	1.53 ± 0.07*

$*p < 0.01$.

Type II cells were cultured in the presence or absence of cortisol (10^{-5} M) as shown schematically in Figure 6. After 44 h of culturing, the indicated radioactive precursors were added to estimate the formation of PC, DSPC and PG. Results are expressed as ratios cortisol-treated/control which were calculated in each separate experiment. The data are averages ± SEM obtained in the indicated number of experiments. Statistical significance was determined by using Student's t test.

phatidylglycerols in fresh type II cells from normal adult rat lung in primary culture. The procedure is shown schematically in figure 6. A partially purified preparation of type II cells was obtained from rat lung by trypsinization and density gradient centrifugation [45, 47]. Further purification was achieved by differential adherence in primary monolayer culture [44, 45]. During the first 3 h of culture most macrophages adhere. The nonadherent cells were subsequently transferred to a series of small (35 mm) tissue culture dishes and culturing continued for 20 h. During this period the type II cells will adhere whereas most other cells do not. At the end of this period the medium and nonadherent cells were replaced by fresh medium containing cortisol (10^{-5} M) and the type II cells cultured for an additional 24 h. At the end of this period the medium was replaced by a serum-free incubation medium containing cortisol (10^{-5} M) and one of the following labelled substrates: 0.025 mM [Me-14C]choline, 1 mM [1-14C]acetate or 0.2 mM [1-14C]palmitate, 5.6 mM [U-14C]glucose or 0.1 mM [1,3-3H]glycerol [Post, Batenburg and Van Golde, in preparation]. After a period of 5 h, the incubations were terminated [9] and the incorporations of the various labelled substrates into phosphatidylcholines and phosphatidylglycerols estimated.

The number of type II cells per dish, the purity and viability (as judged from trypan blue exclusion) were not affected by exposure to cortisol (data not shown). The effects of cortisol on the incorporation of labelled precursors into total and disaturated phosphatidylcholines and into phosphatidylglycerols are shown in table I. Exposure of the type II cells to cortisol had a significant ($p < 0.01$) stimulatory effect upon the entry of all labelled precursors into total and disaturated phosphatidylcholines. The percentage disaturation of phosphatidylcholines synthesized from the various precursors was, however, not significantly influenced by cortisol.

Cortisol also enhanced the synthesis of phosphatidylglycerols from labelled acetate, palmitate, glycerol and glucose. These data corroborate earlier studies with whole lung [2] that cortisol may not only accelerate the formation of surfactant lipids in fetal lung but also be involved in regulating this process in the adult lung. It would be of great interest to measure the activities of the various enzymes involved in the synthesis of surfactant lipids after exposure of the cells to cortisol. Such studies could pinpoint the hormonally regulated steps in these processes.

The present studies focussed on the direct action of cortisol on isolated type II cells. *Smith* [55, 56] demonstrated recently that an oligopeptide factor produced by fetal lung fibroblasts in response to cortisol, stimulated the formation of disaturated phosphatidylcholines in cloned type II cells. This so-called fibroblast-pneumonocyte factor also accelerated fetal lung maturation *in vivo* [56]. *Smith* suggested that the glucocorticosteroid effect on lung maturation might depend upon intercellular interactions. It would be of great interest to investigate whether the direct cortisol stimulation and the oligopeptide-mediated stimulation are due to effects on the same enzymes in the type II cells.

Synthesis of Phosphatidylcholine and Phosphatidylglycerol from Radioactive Acetate and Palmitate: A Comparison between the Isolated Perfused Lung and Isolated Alveolar Type II Cells

Longmore et al. [41] perfused isolated rat lungs with a series of labelled precursors to obtain microsomes which contained sufficiently labelled endogenous phosphatidylcholines to investigate the specificity of microsomal phospholipase A_2 towards membrane-bound substrates. In the course of that study it was found that [9, 10-$^3H_2$]palmitate was incorporated predominantly into the 2-position of both total and disaturated phos-

Table II. Positional distribution of radioactivity from palmitate and acetate in phosphatidylcholines of perfused isolated rat lung and of isolated alveolar type II cells from adult rat lung

Substrate	% of label at 2-position of total phosphatidylcholines			% of label at 2-position of disaturated phosphatidylcholines		
	whole[1] lung	lung[1] surfactant	isolated[2] type II cells	whole[1] lung	lung[1] surfactant	isolated[2] type II cells
Acetate	36 ± 4	65 ± 2	64	33 ± 5	72 ± 1	88
Palmitate	69 ± 2	74 ± 3	63	91 ± 1	80 ± 3	74

The lungs were perfused with trace amounts of [1–^{14}C]acetate (0.0164 mM) or [9,10-^{3}H$_2$]palmitate (0.006 mM) for 1.5 h. The surfactant preparations were obtained by lavage of the lungs after terminating the perfusion period. Isolated type II cells were incubated for 1 h with trace amounts of [1–^{14}C]acetate (0.016 mM) or [1–^{14}C]palmitate (0.008 mM) [for details see 10].

[1] Results are averages $\pm$ SEM of at least 4 experiments.
[2] Data are representative for results obtained in 4 experiments, each carried out in duplicate.

phatidylcholines of whole lung tissue, an observation which agreed with earlier studies *in vivo* [48]. After perfusion of the lungs with *trace* amounts of [1-^{14}C]acetate, however, the radioactivity was recovered mainly at the 1-position of both total and disaturated phosphatidylcholines of whole lung tissue. In addition, microsomal phosphatidylcholines labelled during perfusion with trace amounts of [1-^{14}C]acetate were utilized as substrate by the microsomal phospholipase A$_2$ whereas those labelled during perfusion with trace amounts of [9,10-^{3}H$_2$]palmitate were not degraded by this enzyme, despite the fact that the majority of the [9,10-^{3}H$_2$]palmitate was present at the 2-position. Also experiments *in vivo* by *Jobe* [32] indicated a different utilization of endogenously synthesized and exogenously supplied palmitate in the formation of lung tissue and surfactant lipids. However, both the studies of *Longmore et al.* [41] and those by *Jobe* [32] were performed with whole lung.

In order to further investigate possible differences in metabolic fate between endogenously synthesized and exogenously supplied palmitate, freshly isolated type II cells were incubated in the presence of trace amounts of either [1-^{14}C]acetate or [1-^{14}C]palmitate (table II) [10]. Interestingly, the labels from palmitate and acetate entered both predominantly into the 2-position of total and disaturated phosphatidylcholines of type II cells. The discrepant results obtained with the perfused whole lung on the one hand and those with isolated type II cells on the other hand (table II), suggested that the different positional distribution of [1-^{14}C]acetate observed in the perfused whole lung may reflect processes occurring in other cell types than the type II cells. This suggestion was corroborated by the finding (table II) that in the surfactant fraction prepared from the perfused lung by lavage, the label from [9,10-^{3}H$_2$]palmitate and that from [1-^{14}C]acetate were both recovered mainly at the 2-position as had been observed for the type II cells. Although the perfusions in the present study were carried out for a fixed time, it seems justified from the findings in table II to stress once more that results obtained in experiments with whole lung cannot always be extrapolated to type II cells. However, the data in table II also suggest that the isolated perfused lung may be a good model to study the biochemical processes related to surfactant synthesis going on in the type II cells in their natural environment, provided that in some way the products of the processes occurring in the type II cells are separated from the products of other cell types after the perfusion for example by lung lavage or by isolation of lamellar bodies after the perfusion.

No difference was observed between lung tissue and surfactant with regard to the positional distribution of label from trace amounts of [1-^{14}C]acetate in phosphatidylglycerol (data not shown). This could be explained by assuming that the type II cells, the producers of surfactant, are the only cells in lung tissue that synthesize appreciable amounts of phosphatidylglycerol. This assumption would be in line with the findings that pulmonary surfactant is unique in its high phosphatidylglycerol content [28, 52].

Acknowledgements

The authors are much indebted to Miss *W. Klazinga* and Mr. *V. Oldenborg* for excellent technical assistance and to Mr. *C.J.A.H.V. Van Vorstenbosch* for electron microscopic analyses. The investigations were supported in part by the Netherlands Foundation for Chemical Research (SON) with financial aid from the Netherlands Organization for the Advancement of Pure Research (ZWO). The authors gratefully acknowledge the financial support by Dr. Karl Thomae GmbH, Biberach, for a number of experiments presented in this study.

References

1 Abe, M.; Akino, T., and Ohno, K.: The formation of lecithin from lysolecithin in rat lung supernatant. Biochim. biophys. Acta *280:* 275–280 (1972).

2 Abe, M. and Tierney, D.F.: Lung lipid metabolism after 7 days of hydrocortisone administration to adult rats. J. appl. Physiol. *42:* 202–205 (1977).

3 Anderson, G.G.; Cidlowski, J.A.; Absher, P.M.; Hewitt, J.R., and Douglas, W.H.J.: The effect of dexamethasone and prostaglandin $F_{2\alpha}$ on production and release of surfactant in type II alveolar cells. Prostaglandins *16:* 923–929 (1978).

4 Avery, M.E.: Pharmacological approaches to the acceleration of fetal lung maturation. Br. med. Bull. *31:* 13–17 (1975).

5 Ballard, P.L.; Benson, B.J., and Brehier, A.: Glucocorticoid effects in the fetal lung. Am. Rev. resp. Dis. *115:* 29–35 (1977).

6 Ballard, P.L.; Mason, R.J., and Douglas, W.H.J.: Glucocorticoid binding by isolated lung cells. Endocrinology *102:* 1570–1575 (1978).

7 Ballard, R.A. and Ballard, P.L.: Use of prenatal glucocorticoid therapy to prevent respiratory distress syndrome. A supporting view. Am. J. Dis. Child. *130:* 982–987 (1976).

8 Batenburg, J.J.; Longmore, W.J.; Klazinga, W., and Van Golde, L.M.G.: Lysolecithin acyltransferase and lysolecithin:lysolecithin acyltransferase in adult rat lung alveolar type II epithelial cells. Biochim. biophys. Acta *573:* 136–144 (1979).

9 Batenburg, J.J.; Longmore, W.J., and Van Golde, L.M.G.: The synthesis of phosphatidylcholine by adult rat lung alveolar type II epithelial cells in primary culture. Biochim. biophys. Acta *529:* 160–170 (1978).

10 Batenburg, J.J.; Post, M.; Oldenborg, V., and Van Golde, L.M.G.: The perfused isolated lung as a possible model for the study of lipid synthesis by type II cells in their natural environment. Exp. Lung Res. *1:* 57–65 (1980).

11 Batenburg, J.J. and Van Golde, L.M.G.: Formation of pulmonary surfactant in whole lung and in isolated type II alveolar cells; in Scarpelli and Cosmi, Reviews in perinatal medicine, vol. 3, pp. 73–114 (Raven Press, New York 1979).

12 Brumley, G. and Van Den Bosch, H.: Lysophospholipase-transacylase from rat lung: isolation and partial purification. J. Lipid Res. *18:* 523–532 (1977).

13 Douglas, W.H.J. and Farrell, P.M.: Isolation of cells that retain differentiated functions in vitro: Properties of clonally isolated type II alveolar pneumonocytes. Environ. Health Perspect. *16:* 83–88 (1976).

14 Douglas, W.H.J.; Moorman, G.W., and Teel, R.W.: The formation of histotypic structures from monodisperse fetal rat lung cells cultured on a three-dimensional substrate. In Vitro *12:* 373–381 (1976).

15 Ekelund, L.; Arvidson, G., and Åstedt, B.: Cortisol-induced accumulation of phospholipids in organ culture of human fetal lung. Scand. J. clin. Lab. Invest. *35:* 419–423 (1975).

16 Erbland, J.F. and Marinetti, G.V.: The enzymatic acylation and hydrolysis of lysolecithin. Biochim. biophys. Acta *105:* 128–138 (1965).

17 Farrell, P.M.: Fetal lung development and the influence of glucocorticoids on pulmonary surfactant. J. Steroid Biochem. *8:* 463–470 (1977).

18 Farrell, P.M. and Avery, M.E.: Hyaline membrane disease. Am. Rev. resp. Dis. *111:* 657–688 (1975).

19   Farrell, P.M. and Zachman, R.D.: Induction of choline phosphotransferase and lecithin synthesis in the fetal lung by corticosteroids. Science, N.Y. *179:* 297–298 (1973).

20   Finkelstein, J.N. and Mavis, R.D.: Biochemical characterization of isolated type II alveolar epithelial cells. Fed. Proc. *37:* 1820 (1978).

21   Fisher, A.B.; Huber, G.A.; Furia, L.; Bassett, D., and Rabinowitz, J.L.: Evidence for lipid synthesis by the dihydroxyacetone phosphate pathway in rabbit lung subcellular fractions. J. Lab. clin. Med. *87:* 1033–1040 (1976).

22   Frosolono, M.F.; Slivka, S., and Charms, B.L.: Acyltransferase activities in dog lung microsomes. J. Lipid Res. *12:* 96–103 (1971).

23   Garcia, A.; Newkirk, J.D., and Mavis, R.D.: Lung surfactant synthesis: A Ca^{2+}-dependent microsomal phospholipase A_2 in the lung. Biochem. biophys. Res. Commun. *64:* 128–135 (1975).

24   Goerke, J.: Lung surfactant. Biochim. biophys. Acta *344:* 241–261 (1974).

25   Gross, I.; Wilson, C.M.; Ingleson, L.D., and Rooney, S.A.: Comparison of the effects of dexamethasone and thyroxine on phospholipid synthesis by fetal rat lung in organ culture. Pediat. Res. *13:* 535 (1979).

26   Hallman, M.; Feldman, B.H.; Kirkpatrick, E., and Gluck, L.: Absence of phosphatidyl-glycerol (PG) in respiratory distress syndrome in the newborn. Study of the minor surfactant phospholipids in newborns. Pediat. Res. *11:* 714–720 (1977).

27   Hallman, M. and Gluck, L.: Phosphatidylglycerol in lung microsomes. I. Synthesis in rat lung microsomes. Biochem. biophys. Res. Commun. *60:* 1–7 (1974).

28   Hallman, M. and Gluck, L.: Phosphatidylglycerol in lung surfactant. II. Subcellular distribution and mechanism of biosynthesis *in vitro.* Biochim. biophys. Acta *409:* 172–191 (1975).

29   Hallman, M. and Raivio, K.: Studies on the biosynthesis of disaturated lecithin of the lung: the importance of the lysolecithin pathway. Pediat. Res. *8:* 874–879 (1974).

30   Hamosh, M. and Hamosh, P.: Lipoprotein lipase in rat lung. The effect of fasting. Biochim. biophys. Acta *380:* 132–140 (1975).

31   Havel, R.J.; Felts, J.M., and Van Duyne, C.M.: Formation and fate of endogenous tri-glycerides in blood plasma of rabbits. J. Lipid Res. *3:* 297–308 (1962).

32   Jobe, A.: An *in vivo* comparison of acetate and palmitate as precursors of surfactant phosphatidylcholine. Biochim. biophys. Acta *572:* 404–412 (1979).

33   Johnston, J.M.; Reynolds, G.; Wylie, M.B., and MacDonald, P.C.: The phosphohy-drolase activity in lamellar bodies and its relationship to phosphatidylglycerol and lung surfactant formation. Biochim. biophys. Acta *531:* 65–71 (1978).

34   Kasten, F.H.: Mammalian myocardial cells; in Kruse and Patterson, Tissue culture-methods and applications, pp. 72–81 (Academic Press, New York 1973).

35   Kennedy, E.P.: Biosynthesis of complex lipids. Fed. Proc. *20:* 934–940 (1961).

36   Kikkawa, Y.; Aso, Y.; Yoneda, K., and Smith, F.: Lecithin synthesis by normal and bleomycin-treated type II cells; in Bouhuys, Lung cells in disease, pp. 139–146 (Elsevier, Amsterdam 1976).

37   Kikkawa, Y. and Yoneda, K.: The type II epithelial cell of the lung. I. Method of isola-tion. Lab. Invest. *30:* 76–84 (1974).

38   Kikkawa, Y.; Yoneda, K.; Smith, F.; Packard, B., and Suzuki, K.: The type II epithelial cells of the lung. II Chemical composition and phosopholipid synthesis. Lab. Invest. *32:* 295–302 (1975).

39   King, R.J. and Clements, J.A.: Surface active materials from dog lung. II. Composition and physiological correlations. Am. J. Physiol. *223:* 715–726 (1972).

40   Lands, W.E.M.: Metabolism of glycerolipids. II. The enzymatic acylation of lysolecithin. J. biol. Chem: 2233–2237 (1960).

41   Longmore, W.J.; Oldenborg, V., and Van Golde, L.M.G.: Phospholipase A_2 in rat-lung microsomes: substrate specificity towards endogenous phosphatidylcholines. Biochim. biophys. Acta *572:* 452–460 (1979).

42   Mason, R.J.: Importance of the acyldihydroxyacetone phosphate pathway in the synthesis of phosphatidylglycerol and phosphatidylcholine in alveolar type II cells. J. biol. Chem. *253:* 3367–3370 (1978).

43   Mason, R.J.; Dobbs, L.G.; Greenleaf, R.D., and Williams, M.C.: Alveolar type II cells. Fed. Proc. *36:* 2697–2702 (1977).

44   Mason, R.J. and Williams, M.C.: Type II alveolar cell-defender of the alveolus. Am. Rev. resp. Dis. *115:* 81–91 (1977).

45   Mason, R.J.; Williams, M.C., and Dobbs, L.G.: Secretion of disaturated phosphatidyl-choline by primary cultures of type II alveolar cells; in Sanders, Schneider, Dagle and Ragan, Pulmonary macrophage and epithelial cells. Proc. 16th Annu. Hanford Biol. Symp., Energy Res. Dev. Admin. Symp. Ser. 43, pp. 280–297 (Technical Information Center, Energy research and Development Administration, Springfield, Va. 1977).

46   Mason, R.J.; Williams, M.C., and Greenleaf, R.D.: Isolation of lung cells; in Bouhuys, Lung cells in disease, pp. 39–52 (Elsevier, Amsterdam 1976).

47   Mason, R.J.; Williams, M.C.; Greenleaf, R.D., and Clements, J.A.: Isolation and properties of type II alveolar cells from rat lung. Am. Rev. resp. Dis. *115:* 1015–1026 (1977).

48   Moriya, T. and Kanoh, H.: *In vivo* studies on the *de novo* synthesis of molecular species of rat lung lecithins. Tohoku J. exp. Med. *112:* 241–256 (1974).

49   Ohta, M. and Hasegawa, H.: Phospholipase A activity in rat lung. Tohoku J. exp. Med. *108:* 85–94 (1972).

50   Picken, J.; Lurie, M., and Kleinerman, J.: Mechanical and morphologic effects of long term corticosteroid administration on the rat lung. Am Rev. resp. Dis. *110:* 746–753 (1974).

51   Rooney, S.A.: Biosynthesis of lung surfactant during fetal and early postnatal development. TIBS *4:* 189–191 (1979).

52   Rooney, S.A.; Canavan, P.M., and Motoyama, E.K.: The identification of phosphat-idylglycerol in the rat, rabbit, monkey and human lung. Biochim. biophys. Acta *360:* 56–67 (1974).

53   Russell, B.J.; Nugent, L., and Chernick, V.: Effects of steroids on the enzymatic pathways of lecithin production in fetal rabbits. Biol. Neonate *24:* 306–314 (1974).

54   Smith, B.T.: Cell line A 549: A model system for the study of alveolar type II cell function. Am. Rev. resp. Dis. *115:* 285–293 (1977).

55   Smith, B.T.: Fibroblast-pneumonocyte factor: Intercellular mediator of glucocorticoid effect on fetal lung; in Stern, Neonatal intensive care, pp. 25–32 (Masson, Boston 1978).

56   Smith, B.T.: Lung maturation in the fetal rat: acceleration by injection of fibroblast-pneumonocyte factor. Science, N.Y. *204:* 1094–1095 (1979).

57   Smith, B.T. and Torday, J.S.: Factors affecting lecithin synthesis by fetal lung cells in culture. Pediat. Res. *8:* 848–851 (1974).

58   Smith, F.B. and Kikkawa, Y.: The type II epithelial cells of the lung. III. Lecithin synthesis: a comparison with pulmonary macrophages. Lab. Invest. *38:* 45–51 (1978).

59   Snyder, C.; Malone, B.; Nettesheim, P., and Snyder, F.: Urethan-induced pulmonary adenoma as a tool for the study of surfactant biosynthesis. Cancer Res. *33:* 2437–2443 (1973).

60   Sorokin, S.P.: The cells of the lung; in Nettesheim Hanna and Deatherage, Morphology of experimental respiratory carcinogenesis, pp. 3–43 (US Atomic Energy Commission, Oak Ridge 1970).

61   Tansey, F.A. and Frosolono, M.F.: Role of 1-acyl-2-lyso-phosphatidylcholine acyl transferase in the biosynthesis of pulmonary phosphatidylcholine. Biochim. biophys. Res. Commun. *67:* 1560–1566 (1975).

62   Van Golde, L.M.G.: Metabolism of phospholipids in the lung. Am. Rev. resp. Dis. *114:* 977–1000 (1976).

63   Vereyken, J.M.; Montfoort, A., and Van Golde, L.M.G.: Some studies on the biosynthesis of molecular species of phosphatidylcholine from rat lung and phosphatidylcholine and phosphatidylethanolamine from rat liver. Biochim. biophys. Acta *260:* 70–81 (1972).

64   Wykle, R.L., Malone, B., and Snyder, F.: Biosynthesis of dipalmitoyl-*sn*-glycero-3-phosphocholine by adenoma alveolar type II cells. Archs Biochem. Biophys. *181:* 249–256 (1977).

L.M.G. Van Golde, Ph.D, Laboratory of Veterinary Biochemistry, State University of Utrecht, Bilstraat 172, 3572 BP Utrecht (The Netherlands)

Discussion

Van Golde, in reply to a question of *Tierney,* reported that they did not use lyso-PC in incubations with intact type II cells because it is possible that this substrate is degraded by a lysophospholipase in the cell membrane of the type II cell. Lysophospholipase occurs in the membrane of many cell types, possibly also in that of the type II cell. Therefore it is difficult to judge whether the lyso-PC enters into the cell as an intact molecule. Recent studies of *Van den Bosch* with mixtures of D- and L-lyso-PC have shown that there is hardly any formation of dipalmitoylphosphatidylcholine in the adult rat lung *in vivo.* These observations seem to confirm the findings with isolated type II cells that the enzyme catalyzing the transacylation is much less active than the acylation activity. In fetal lung it may be different.

Rüfer found an increase in lysolecithin in experiments with shock and asked about the toxicity of lysolecithins. *Van Golde* agreed that high levels of lysolecithin may be toxic to the type II cells. The concentrations required for the remodeling are, however, very small.

Van Golde, referring to a question by *Von Wichert* said that if lysolecithin levels increase in pathological conditions, the activity of the transacylation process might become more important.

Clements inquired about any significant remodeling of phosphatidylcholine components of surfactant in lamellar bodies, to which *Van Golde* replied that there is not much evidence for significant remodeling in lamellar bodies. This view is supported by recent findings of *Voelker* and *Snyder* who determined the positional distribution of radioactive palmitate between the 1- and 2-position of phosphatidylcholine in microsomes and lamellar bodies of pulmonary adenoma as a function of time. Also their data suggest that the remodeling proceeds in the microsomes.

Van Golde, answering a question of *Von Wichert,* thought there might be species differences in lung phospholipid metabolism, particularly in fetal lung. In adult lung the data so far available, give little indication of species differences in pulmonary lipid metabolism.

Prog. Resp. Res., vol. 15, pp. 20–26 (Karger, Basel 1981)

Secretion and Clearance of Lung Surfactant: A Brief Review[1]

John A. Clements[2], Manuel J. Oyarzún and Aldo Baritussio

Cardiovascular Research Institute, University of California, San Francisco, Calif., Departamento de Medicina Experimental, Universidad de Chile, Santiago de Chile, and Department of Medicine, University of Padova, Padova

In discussing the functions of lung surfactant, it is usual to emphasize its ability to form an interfacial film and thus reduce the surface tension of the pulmonary airspaces to very low values. It is less common to describe the metastable character of the surfactant film, its inevitable loss from the interface when lung volume (i.e., area) is decreased, and the resulting need for continual turnover of film components. Although one does not know at present what fraction of the film that is lost during reduction of volume can reenter the interface with adequate lung expansion, the prominent pressure-volume hysteresis that accompanies reinflation from low volumes [1] and the progressive loss of compliance and atelectasis that occur during shallow breathing [2, 3] suggest that this fraction is significantly less than 100%. If this be true, it follows that maintenance of an adequate reserve of functional surfactant in the alveolar subphase requires ongoing secretion from intracellular sites of synthesis and storage. It is also obvious that surfactant components must normally be removed from the alveoli at a comparable rate, since they are not continually accumulated or depleted. Further, since the movements of the components of the film are influenced by changes in lung volume and thus by the pattern of breathing, it is necessary that the processes of synthesis, secretion, and clearance of the components respond appropriately to ventilatory changes, if the pool of instantly available survactant is to be properly regulated [4].

What are the pathways of surfactant flux into and out of the airspaces? How are the rates of inflow and removal keyed to ventilatory require-

[1] Supported in part by Grant HL-06285 from the National Institutes of Health.
[2] Career Investigator, American Heart Association.

ments? What control signals are generated and how are they transduced? Can the physician modify these processes to benefit patients with lung disease? The work summarized in this paper was designed to answer some of these questions. Details of the biosynthesis of surfactant components and the role of alveolar epithelial type II cells in these processes have been extensively reviewed [5, 6] and will not be repeated here.

Methods

Lung surfactant is a phospholipid-rich, protein-containing complex [7] whose principal component, saturated phosphatidylcholine (SPC), appears normally to constitute most of the stabilizing interfacial film [4, 8] at functional residual volume. It is convenient to observe changes in extracellular surfactant by measuring the amounts of SPC and phospholipid (PL) in lung lavage liquid, and this is the method we have used in studying how the surfactant pool is regulated.

We have described the methods for these studies in detail [9, 10]. Briefly, rabbits weighing 2–4 kg were anesthetized with 30 mg/kg i.v. sodium pentobarbital, supplemented as needed to maintain light anesthesia. The trachea was cannulated for registration of air flow, tidal volume, respiratory rate and minute ventilation. A femoral artery was cannulated for measurement of blood pressure and arterial blood pH, PCO_2, and PO_2. After a control period of 1 h of spontaneous breathing an experimental intervention was made, lasting from 15 min to 4 h . Then, following a lethal dose of pentobarbital, the lungs were removed, separated into left and right, degassed and weighed. The lungs were lavaged 3 times with maximal expansion, using room temperature isotonic sodium chloride solution. The washings were pooled for analysis. Lipids were extracted and total PL and SPC determined. Results were expressed as milligram PL/gram fresh lung. Experiments were accepted only if arterial pH and gas tensions remained within control levels; lung appearance was normal; lung weight/body weight and wet/dry weight ratios were within control ranges; and lavage liquid protein concentration, lactate dehydrogenase activity, and cell counts were within control ranges (mean ± 1 SD). To check whether the experimental manipulations caused contamination of lavage liquid with PL other than surfactant PL, we determined the ratio of SPC/PL in 126 experiments. No condition produced a significant deviation of this ratio from the value (0.57) characteristic of purified surfactant [7]. During the experimental period we either added enough deadspace to the tracheal cannula to double minute ventilation, administered a drug, stimulated the distal end of the cut left cervical vagus nerve, or combined one of these stimuli with administration of an antagonist drug. In recent experiments we placed radioactively labelled *L*- or *D*-dipalmitoyl phosphatidylcholine isomers (*L*-DPPC and *D*-DPPC) in small unilamellar liposomes onto the alveolar surfaces, using as a vehicle a small volume of ½ strength Ringer's solution, and followed their clearance for periods up to 5 h . The effects of doubling minute ventilation or adding 20% by weight of the fusogenic lipid phosphatidylglycerol on clearance rates were determined. We also measured the rates at which the labelled DPPC in these liposomes could be hydrolyzed by homogenized lung tissue and snake venom phospholipases. Detailed descriptions of these experiments have been submitted for publication [11].

Results

Increased ventilation increased the extracellular pool of PL by 45% at 1 h , 54% at 2 h , and 25% at 4 h . At 2 h pool size correlated significantly with tidal volume but not with respiratory frequency, minute ventilation, or arterial blood gas levels or pH. The principal determinant of pool size thus seemed to be how much the lungs were stretched during breathing (the 'stretch effect').

To try to find out how stretching the lungs might give a signal for increasing the extracellular PL pool, we administered blocking agents just before and during increased ventilation. The beta-adrenergic blockers propranalol and sotalol both completely prevented the expected increase in PL. The alpha blocker phenoxybenzamine (at a dose which completely blocked the cardiovascular effects of a large bolus of norepinephrine) failed entirely to reduce the response of PL pool size to increased ventilation.

Because adrenergic stimuli can sometimes be potentiated by certain prostanoids and because distortion of tissues can lead to release of such substances, we attempted to block the stretch effect with the prostaglandin synthetase inhibitors indomethacin and sodium meclofenamate. Both blocked the effect completely.

Because of many reports [summarized in 9] that the cholinergic system may regulate surfactant secretion, we examined the effects of administering atropine or cooling the cervical vagi to 3.5 °C. Atropine sulfate completely blocked the stretch effect without altering the size of the PL pool in control animals or changing ventilatory pattern or blood gases. Cooling the vagi only partially prevented the effect.

Since blocking agents interfered with the stretch effect, we attempted to imitate it with agonists. Infusion of acetylcholine into one lung caused only a modest (13%) increase in the PL pool. Stimulation of the distal end of the cut left cervical vagus did not produce a unilateral effect, but it did raise the pool 31% bilaterally in stimulated animals as compared to controls. Infusion of prostaglandins E_1, E_2, and $F_{2\alpha}$ did not cause significant changes in pool size, whereas the stable endoperoxide analog (15s) hydroxy 11α, 9α-(epoxymethano) prosta-5Z, 13E-dienoic acid (Upjohn 44619) caused a 37% fall in only 15 min. On the other hand, infusion of the β_2-agonist terbutaline sulfate raised the pool substantially (40%) without causing significant changes in respiration, blood gases or cardiovascular measurements. Propranalol completely blocked the effect of terbutaline on the PL pool size.

We conclude from these results that the regulation of the extracellular pool of PL in the lung is complex. The evidence seems clear that when breathing increases so does the surfactant available in airspaces and that this effect depends mainly on increase in tidal volume. Cholinergic, β-adrenergic, and prostaglandin synthesis blockers can each completely prevent this response, but the agonists vary widely in their ability to reproduce it under the conditions of these experiments. $β_2$ stimulation is very effective, acetylcholine and vagal stimulation are less so, and E and $F_{2α}$prostaglandins are ineffective. The sharp decrease in PL pool with the cyclic endoperoxide analog U 44619 remains to be explained.

These experiments do not tell to what extent control of the PL pool is accomplished within the lungs or how much other organs are involved. Nor can they separate out the extent to which a given change in the PL pool reflects a change in secretion, a change in clearance, or both. Experiments with labelling of surfactant components are needed to quantify these reactions. The relative importance of the several mediator systems *in vivo* needs further clarification. It is helpful in designing such studies to know (1) that $β_2$ agonists and dibutyryl cyclic AMP very strongly stimulate secretion of SPC from rat isolated type II cells, but cholinergic agonists and dibutyryl cyclic GMP do not [12], and (2) that increasing the tidal volume in artificially ventilated rats raises the specific activity of labelled SPC in lavage fluid and elevates cyclic GMP (but not cyclic AMP) in the lung tissue [13].

To what extent is clearance of PL from the airspaces affected by ventilation of the lungs and by the composition of the PL? Are PL degraded in the airspaces or removed intact? To obtain partial answers to these questions we placed radioactively labelled DPPC in 40–60 nm DPPC liposomes on the airspace surfaces of rabbits and followed its disappearance as measured in lavage fluid [11].

When the liposomes contained [14]C-*L*-DPPC, the label disappeared from the airspaces at an average rate of 7.8%/h during normal breathing. Adding enough deadspace to the tracheal cannula to double minute ventilation increased the average rate of clearance to 13.3%/h. About 95% of the label was in *L*-DPPC, whether it was recovered from lavage liquid or lung tissue. When the liposomes were prepared with 20% by weight of phosphatidylglycerol and administered to normally breathing animals the labelled *L*-DPPC was cleared at a rate of 13.9%/h.

We also had added [3]H-*D*-DPPC, the dextro isomer, to the liposomes used in these experiments because it was expected to resist enzymatic degradation, and thus its clearance would have to occur by nonenzymatic processes. The ^{3}H-D-DPPC disappeared from lavage fluid at the same rate as ^{14}C-L-DPPC. When we determined the amounts of the labelled isomers remaining in the whole lung (tissue plus airspaces), we found most ($>90\%$) of the labels still in DPPC, and both disappeared at a rate of about 3%/h. We never found more than 1% of the labels in the trachea even after 5 h when 60–70% had disappeared from the airspaces, but 50–60% remained in the lungs. Less than 3% of lavage PL and DPPC are in alveolar macrophages [10]. Incubation of the liposomes with homogenized lung tissue resulted in hydrolysis of the L-DPPC at a rate of 2.1%/h, but the D-DPPC was not attacked at all. Snake venom phospholipase A_2 hydrolyzed the L-DPPC in these liposomes very rapidly but did not split the D-DPPC.

In these experiments the liposomes were made principally of L-DPPC and the labelled D and L isomers were present only in trace amounts. The whole mass administered was calculated to be about 10% of the DPPC already present on the airspace surfaces. Judging from respiratory and circulatory measurements, blood gases, lavage protein, LDH, and cell count, lung weight/body weight ratio, lung appearance on gross, light microscopic, and electron microscopic examination, the liposomes were well distributed in the alveoli, the vehicle was quickly absorbed, the injected lung was not damaged, and the animal was not greatly disturbed.

We conclude that under the conditions of these studies, exogenous DPPC is cleared from airspaces by nonenzymatic processes deep in the lungs; that a negligible part reaches the upper airway; that when the natural L-isomer penetrates the tissue its net hydrolysis is slow; that increasing lung ventilation speeds DPPC clearance from the airspaces (but not the tissue); and that addition of the fusogenic lipid phosphatidylglycerol accelerates DPPC loss. In these experiments DPPC was presented in a form (small unilamellar liposomes) apparently different from normal intraalveolar surfactant, and we cannot be sure from these studies that the DPPC was cleared through normal channels and at normal rates. The information we have obtained should be useful, nevertheless, in designing artificial surfactants for replacement therapy in diseases characterized by surfactant deficiency.

The results of all our experiments on control of surfactant pool size in airspaces and on clearance of DPPC should be compared with tracer studies of the turnover of surfactant components [13–16]. The speeds at which intravenously administered labelled precursors begin to appear in airspace PL and DPPC (about 1 h) contrast sharply with the prolonged residence of the labels in the extracellular pools (up to 2 days). In one paper [17] it was reported that the label in ^{3}H-DPPC given as an aerosol had largely entered the alveolar epithelium within an hour. Our experiments show that the pool size can be abruptly shifted by drugs or changes in lung ventilation and that clearance of administered DPPC can be accelerated by increased breathing. The most likely interpretation of all these findings is that secretion and clearance of the components of the extracellular surfactant are under minute-to-minute control and that the components largely reenter the epithelium and at least some moieties are reutilized. Presumably these processes express 'the wisdom of the body', and physicians may someday learn how to manipulate them for the benefit of their patients.

References

1   Radford, E.P., Jr.: Static mechanical properties of mammalian lungs; in Fenn and Rahn, Handbook of physiology, section 3, vol. 1, pp. 429–449 (American Physiological Society, Washington, D.C., 1964).

2   Mead, J. and Collier, C.: Relation of volume history of lungs to respiratory mechanics in anesthetized dogs. J. appl. Physiol. *14*: 669–678 (1959).

3   Thet, L.A.; Clerch, L.: Massaro, G.D., and Massaro, D.: Changes in sedimentation of surfactant in ventilated excised rat lungs. J. clin. Invest. *64*: 600–608 (1979).

4   Clements, J.A.: Kinetic properties of lung surfactant. J. Jap. med. Soc. Biol. Interface *8*: 1–7 (1977).

5   Golde, L.M.G., van: Metabolism of phospholipids in the lung. Am. Rev. resp. Dis. *114*: 977–1000 (1976).

6   Batenburg, J.J. and Golde, L.M.G., van: Formation of pulmonary surfactant in whole lung and in isolated type II alveolar cells; in Scarpelli and Cosmi, Reviews in perinatal medicine, vol. 3, pp. 73–114 (Raven Press, New York 1979).

7   King, R.J. and Clements, J.A.: Surface active materials from dog lung. II. Composition and physiological correlations. Am. J. Physiol. *223*: 715–726 (1972).

8   Hildebran, J.N.; Goerke, J., and Clements, J.A.: Pulmonary surface film stability and composition. J. appl. Physiol: Respirat. Environ. Exercise Physiol. *47*: 604–611 (1979).

9   Oyarzún, M.J. and Clements, J.A.: Ventilatory and cholinergic control of pulmonary surfactant in the rabbit. J. appl. Physiol. *43*: 39-45 (1977).

10 Oyarzún, M.J. and Clements, J.A.: Control of lung surfactant by ventilation, adrenergic mediators, and prostaglandins in the rabbit. Am Rev. resp. Dis. *117*: 879–891 (1978).

11 Oyarzún, M.J.; Clements, J.A., and Baritussio, A.: Ventilation enhances pulmonary alveolar clearance of radioactive dipalmitoyl phosphatidylcholine in liposomes (Am. Rev. resp. Dis. submitted in 1979).

12 Dobbs, L.G. and Mason, R.J.: Pulmonary alveolar type II cells isolated from rats. Release of phosphatidylcholine in response to β-adrenergic stimulation. J. clin. Invest. *63*: 378–387 (1979).

13 Klass, D.J.: Dibutyryl cyclic GMP and hyperventilation promote rat lung phospholipid release. J. appl. Physiol.: Respirat. Environ. Exercise Physiol. *47:* 285–289 (1979).

14 Young, S.L. and Tierney, D.F.: Dipalmitoyl lecithin secretion and metabolism by the rat lung. Am. J. Physiol. *222:* 1539–1544 (1972).

15 King, R.J.; Martin, H.; Mitts, D., and Holmstrom, F.M.: Metabolism of the apoproteins in pulmonary surfactant. J. appl. Physiol.: Respirat. Environ. Exercise Physiol. *42:* 483–491 (1977).

16 Jobe, A.: The labelling and biological half-life of phosphatidylcholine in subcellular fractions of rabbit lung. Biochim. biophys. Acta *489:* 440–453 (1977).

17 Geiger, K.; Gallagher, M.L., and Hedley-White, J.: Cellular distribution and clearance of aerosolized dipalmitoyl lecithin. J. appl. Physiol. *39:* 759–766 (1975).

J.A. Clements, MD, Cardiovascular Research Institute, 1315 Moffitt, University of California, San Francisco, CA 94143 (USA)

Discussion

Van Golde asked *Clements* how unsaturated phosphatidylcholines affect the clearance, but this is, as *Clements* answered, not yet known for certain. *Clements,* asked by *Geiger,* if he used small unilamellar liposomes because he wanted to define the form of the substrate for clearance, also stated that it might be different from the surface material itself. The methods he has used to isolate the material by lavage would collect both the intrinsic natural surface-active material and the liposomes together. To what extent the labelled material may have left the liposomes and have gone into surfactant components and been cleared is unknown. *Clements* stated that the surface-active system is complex and must be simplified for experiments by at least defining some form of lipid which can be presented to the alveolar surface.

Clements referring to a question by *Reiss,* answered that he hadn't made attempts to subfractionate lavage liquid in his experiments, especially not with respect to tubular myelin. He believed that tubular myelin is on the way of extension of the material from the lamellar bodies to sheets at the interface as was suggested in fracture studies. Lung volume has no effect on recovery of surfactant provided that there is an adequate number of lavages, *Clements* stated in answering a question by *Mitzner.*

Tierney wondered if it appeared likely that the rate of loss of surfactant from the alveolar surface and the rate of secretion could ever deplete the lung of surfactant over a period of time for instance by rapid deep ventilations. *Clements,* referring to *Mitzner*'s work, answered this question with yes, especially that it is possible to deplete the alveolar stores by chronic high tidal volume respiration.

Prog. Resp. Res., vol. 15, pp. 27–40 (Karger, Basel 1981)

Fetal Development of Surfactant: Considerations of Phosphatidylcholine, Phosphatidylinositol, and Phosphatidylglycerol Formation

Mikko Hallman

University of California, San Diego, La Jolla, Calif.

Introduction

Recent interest in studies of lung maturation is largely based upon clinical implications, namely to unravel the etiology of respiratory distress syndrome (RDS) and to prevent this major disease in the newborn. In the late 1950, *Avery and Mead* [3] found that the victims of RDS hat deficient surface activity, and proposed, based upon pioneer investigations of *Clements et al.* [6a] and *Pattle* [26], that atelectasis was due to lack of surfactant. In 1971, *Gluck et al.* [10] showed evidence that RDS is a disease of development, caused by immaturity of the surfactant system. These findings have boosted research in this area during the recent years.

Investigations in embryology have disclosed that lung develops as an invagination of the foregut at the end of the 4th week of human gestation. The primitive airways and the corresponding vasculature rapidly proliferate within the surrounding mesenchymal tissue, so that by the end of the 16th week, 15–25 bronchial branches are complete. Morphologically, the lung may be considered as mature enough to support air breathing by 24–25 weeks, when the newly formed respiratory bronchiole are closely enough lined with a capillary network.

Pool Sizes of Disaturated Phosphatidylcholine.

Disaturated phosphatidylcholine (DPC), almost exclusively consisting of dipalmitoyl PC, is the major surface active phospholipid, that is responsible for lowering the surface tension of alveolar interspaces to virtually zero [6b]. Earlier it was even considered to be the surfactant.

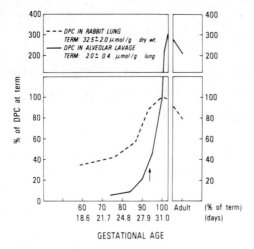

Fig. 1. The mean concentrations of disaturated phosphatidylcholine in lung homogenate and alveolar lavage fluid as studied in the rabbit. The arrow shows the mean gestational age for functional lung maturity.

Numerous investigations have been carried out to clarify the role of DPC as the rate-limiting component in surfactant formation [5a]. Figure 1 shows the concentrations of this phospholipid as measured in lung homogenate and in endobronchial lavage fluid obtained from the rabbit during perinatal development.

DPC in lung parenchyma doubles in concentration during the last fetal week. On the other hand, DPC recovered from alveolar lavage increases more than fivefold during the 3 last fetal days, and consists of about 6% of total lung DPC at term. The birth is accompanied by a further surge of intra-alveolar DPC [11b, 27].

Adams and Fujiwara [2] found that the future airspaces of the fetus are filled with liquid that is secreted by the fetal lung. *Gluck et al.* [10] further confirmed these findings and showed that the surface active phospholipids accumulate in the amniotic fluid. It thus serves as an extracorporeal surfactant storage, and allows the monitoring of surfactant secretions without direct intervention of the fetus.

Figure 2 shows that also in humas, the DPC secretion is preceeded by its parenchymal accumulation. On average, the human lungs are considered to be functionally mature by 88% of the gestational period, whereas rodents become mature only by 94% of term.

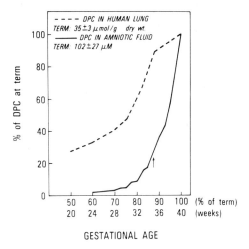

Fig. 2. The mean concentrations of disaturated phosphatidylcholine in human amniotic fluid and in fetal lung homogenate. The arrow shows the mean gestational age for functional lung maturity.

The question why DPC accumulates within the lung cells earlier than it is secreted, is somewhat open. This may be due to failure to secrete the intracellular surfactant storage, failure to form the proper surfactant storage, or accumulation of DPC that is unrelated to the surfactant system.

Besides the changes in quantity, the composition of lung effluent is markedly altered during development. Table I illustrates the lipid composition of alveolar lavage fluid from immature fetuses and that of surfactant isolated from the alveolar lavage of the newborn. In addition to the increase in DPC, another unique surfactant component, phosphatidylglycerol (PG), appears during maturation. On the other hand, other phospholipids, such as sphingomyelin and phosphatidylserine, are prominent in immature lung effluent and almost disappear during development.

The Amniotic Fluid Lung Profile.

Gluck et al. [10], introduced the index of lung maturity, the lecithin/sphingomyelin (L/S) ratio. In this method, DPC is concentrated by precipitation in cold acetone. The content of DPC is expressed on the basis of the reference phospholipid, sphingomyelin. This efficiently comp-

Table I. Lipid composition of alveolar lavage fluid from 27-day-old rabbit fetuses, and of surfactant from newborns

	Immature fetus		Newborn	
	μmol	%	μmol	%
Phospholipid	1.0	100.0	11.1	100.0
Phosphatidylcholine	0.49	49.3	8.85	79.7
DPC	0.14		5.72	
Phosphatidylglycerol	0.00	0.3	0.64	5.8
Phosphatidylinositol	0.06	5.9	0.67	6.0
bis-(Monoacylglycerol)phosphate	0.01	0.9	0.14	1.3
Phosphatidylethanolamine	0.13	13.3	0.59	5.4
Sphingomyelin	0.20	20.5	0.11	1.0
Phosphatidylserine	0.10	9.8	0.09	0.8
'Neutral lipid'	1.2 mg		0.8 mg	
Protein	1.0 mg		1.0 mg	

ensates the fluctuation in DPC concentration, caused by changes in amniotic fluid volume. At 35 ± 2 weeks of gestation, the L/S ratio reaches the value of two, indicating that the lungs are functionally mature.

In the majority of cases, the L/S ratio is a reliable index of maturity. However, transitional L/S ratios, ranging from 1 to 1.9, predict RDS with an accuracy as low as 60%, indicating that about 40% of these L/S ratios may be noninformative. On the other hand, mature L/S ratios predict the absence of RDS with close to 100% accuracy, except in maternal diabetes, a condition associated with up to 15% of false mature L/S ratios.

These limitations prompted further study of the ontogeny of the minor surfactant phospholipids. *Hallman et al.* [15] showed that phosphatidyl-inositol (PI) and PG developed differently from DPC or the L/S ratio. In human fetuses, PI increases up to 36–37 weeks, slightly preceeding the increase of DPC. PG first appears at the 35th to 37th week of gestation. At the same time the percentage of PI levels and decreases towards the term. These parameters have great potential for index of maturity for the following reasons: PI increases early, indicating ongoing maturation, but excluding maturity in the absence of PG. Since absence of PG in lung effluent of the newborn is a remarkably specific biochemical indicator of RDS [12], just the presence of PG could indicate mature lungs. Frequent contaminants of the amniotic fluid such as blood, and meconium, interfere with practically all lung maturity measurements. However, PG is uniquely concentrated in surfactant, and thus could serve as a sensitive index regardless of contamination.

Further studies seem to verify the previous suggestions. Measurements of the L/S ratio, percent DPC, percent PG and percent PI together, the so-called 'lung profile', increase the accuracy of the immature L/S ratio [23]. This means that with immature L/S ratio and in the absence of PG, the risk of RDS is up to 90%, whereas the presence of PG indicates that the risk of RDS is minimal despite the immature L/S ratio.

The lung profile appears particularly useful in high risk pregnancies. In these instances the maturation may be delayed or accelerated [11a], indicating that the L/S ratio reaches the value of two either after the 37th or before the 33rd week of gestation, respectively. During normal development the various indices of lung maturity bear a characteristic relation to each other. The relationship is best analyzed as a function of the L/S ratio. In high risk pregnancies, the lung profile may be abnormal. The PG/PI ratio as a function of L/S ratio was recently studied in so-called normal and diabetic pregnancies [17]. In diabetes, the median PG/PI ratio was significantly lower. This was due to delay in PG appearance, and due to excessively high PI.

It has recently been shown that the problem of false mature L/S ratio in diabetes may be eliminated when the delivery is delayed until PG appears [7, 17, 22].

Acidic Phospholipids and Surface Activity

The epidemiological evidence presented above suggests that PG improves surfactant function.

Reduction of surface forces seems particularly critical at birth, when the fluid within the airways is replaced with air. This transition expectedly is most problematic with small true alveoli that develop at term. In order to further clarify this question, surface properties and chemical composition of surfactant from immature fetuses and newborn animals were compared. Detailed study on lipids and proteins revealed that only the acidic phospholipids were different, namely, the fetus had no PG, whereas PI was higher than in the newborn. Both the immature and the mature surfactants were able to decrease the surface tension below 10 dyn/cm.

However, the compressibility of surfactant containing PG was lower at low surface tension (8–15 dyn/cm) than that without PG [13]. According to *Clements et al.* [6a, b] low compressibility contributes to the stability of alveoli, especially at low functional residual capacity.

Recent studies further indicate that the acidic phospholipids, particularly PG, improve the surface characteristics of dipalmitoyl PC: This has been documented by applying the artificial surfactant to the airways, deficient in surface-active material [19], by repeatedly compressing the film containing DPC and PG to close to zero dyn/cm on surface balance [4], and by effective adsorption of the disaturated phospholipids from subphase to surface [24].

Biochemical Pathways.

In this brief discussion, the hormonal regulation of surfactant ontogeny will not be reviewed. Instead, the pathways of the PC, PG, and PI synthesis, in relation to fetal development, will be summarized. Unraveling of the particular mechanism of biosynthesis, the rate-limiting steps in particular, serves as an important background for the understanding of hormonal regulation. Various difficulties are involved in studies of the effects of hormones [5a, 8]. Equivocal results may sometimes be due to absence of evidence of rate-limiting steps of biosynthesis, due to absence of proper assay techniques, and in absence of simultaneous documentation of changes in surfactant pools.

Immature alveolar epithelial cells, prior to acquiring characteristic features of type II alveolar cells, contain abundant glycogen storage. Golgi apparatus is identified, too, indicating potential for secretory function. Appearance of lamellar inclusion bodies is accompanied by decrease in glycogen [21]. However, a fair number of lamellar bodies are seen even before the functional maturation.

Glycogen serves as a precursor for both glycerol and fatty acid components of phospholipids. The regulation of glycogen breakdown has been extensively studied using muscle and liver.

According to present evidence, rate-limiting enzymes of glycolysis, glycogen phosphorylase, and phosphofructokinase, are inhibited by ATP and stimulated by ADP. The regulation in the lung may be similar to other organs. If so, activation of phospholipid synthesis could increase energy expenditure, and thus stimulate the breakdown of glycogen.

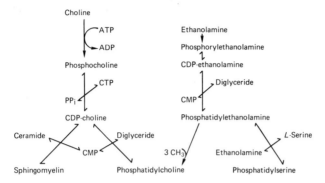

Fig. 3. Biochemical pathways for *de novo* synthesis of phosphatidylcholine, sphingo-myelin, phosphatidylethanolamine and phosphatidylserine.

In addition to glycogen, serum is a continuous source of nutrients, such as glucose, fatty acids, and glycerol. The fetus is mostly dependent on glucose for an energy source, because of active uptake of glucose and relative discrimination of fatty acids by placenta. Accordingly, the fetal lung has a high glucose uptake, and active lipogenesis. Availability of some of the lipid precursors or of lipoprotein lipase in lung endothelium [18], may prove to be important in controlling surfactant production.

Phosphatidic acid phosphohydrolase catalyzes the formation of digly-cerides (*sn*-diacylglycerols) that serve as a common precursor for neutral phospholipids and phosphatidylserine. This activity is associated with various subcellular fractions of the lung, including lamellar bodies and extracellular surfactant. Phosphatidic acid phosphohydrolase in micro-somes and lung supernatant revealed a sharp increase during the last fetal days in rabbit lung [28]. Whether all the activity measured represents surfactant, or even phospholipid synthesis in general, remains in doubt [30].

Whatsoever, phosphatidic acid phosphohydrolase may prove to be important in controlling the availability of diglyceride, but cannot explain the specific increase in PC. Therefore, it is likely that pathways providing choline play an important role in controlling surfactant PC biosynthesis.

Figure 3 illustrates the pathways involved in *de novo* synthesis of PC, sphingomyelin, phosphatidylethanolamine, and phosphatidylserine. The choline incorporation pathway is the main route for surfactant PC forma-tion, whereas the methylation pathway has little importance in this respect

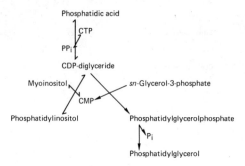

Fig. 4. Biochemical pathways for biosynthesis of phosphatidylinositol and phosphatidylglycerol.

[5a, 8]. The first enzyme of the choline incorporation pathway, choline kinase, catalyzing the formation of phosphocholine, is not rate limiting.

The second enzyme, CDP-choline cytidyltransferase, catalyzes the formation of CDP-choline from phosphocholine. This activity markedly increases during the last fetal days in rat lung. However, addition of lipids restores fetal activity to adult level. Further study of activators revealed that only the acidic phospholipids were effective. The highest increase was with PG, followed by PI, cardiolipin, and phosphatidylserine [9, 29].

The third enzyme, choline phosphotransferase, catalyzes the formation of PC from CDP-choline and diglyceride. Its possible role as a rate-limiting step in PC formation has been extensively studied. However, the suggestion by *Farrel and Morgan* [8], that choline phosphotransferase is rate limiting in *de novo* surfactant PC formation in the fetus has not been substantiated.

The present data indicate that choline phosphotransferase discriminates against disaturated diglycerides. Therefore, DPC seems to be formed through hydrolysis of unsaturated fatty acids at 2-position, followed by reacylation of lysolecithin thus formed. Direct acylation of lysolecithin or transacylation between two lysolecithin molecules serves as a final step of DPC formation. In isolated type II cells, the direct acylation is the predominant activity involved. However, at the moment the quantitative importance of the different saturation mechanisms during fetal development is somewhat open [1, 5a, 16, 25].

Figure 4 shows the pathways for PG and PI formation. CDP-diglyceride serves as a precursor for both these acidic phospholipids. PI formation is catalyzed by CDP-diglyceride: inositol phosphotransferase, whereas PG is formed via two enzymatic steps: glycerophosphate phosphatidyltransferase, and phosphatidylglycerophosphate phosphatase. As compared to glycerophosphate phosphatidyltransferase, phosphatidylglycerophosphate phosphatase is at least 70 times higher in activity. Thus, the former enzyme is likely to be rate limiting.

A recent study indicates that phosphatidate cytidyltransferase activity in microsomes, catalyzing the formation of CDP-diglyceride, triples in activity during the last 5 fetal days and further doubles in activity at birth, as studied in rabbit lung [14].

Increasing microsomal phosphatidate cytidyltransferase expectedly increases CDP-diglyceride, that serves as a precursor for PI and PG. The acidic phospholipid, presumably PI, thus formed may then activate CDP-choline cytidyltransferase by polymerizing the lung supernatant associated enzyme [9]. This could be an important, if not the principal, mechanism to enhance surfactant PC biosynthesis in the fetal lung.

In order to test the hypothesis that the successive development of surfactant PI and PG are controlled by changes in appropriate enzymes, CDP-diglyceride: inositol transferase and glycerophosphate phosphatidyltransferase activities were measured. In lung microsomes CDP: diglyceride-inositol transferase increased during the last fetal days, followed by an increase in glycerophosphate phosphatidyltransferase activity at birth. However, the changes in the respective enzymes were far too modest in order to account for the dramatic changes in surfactant PG and PI [14].

Role of Myoinositol in Regulating the Acidic Surfactant Phospholipids.

As shown in the previous discussion, the development of surfactant PG and PI pools does not satisfactorily correlate with that of the biosynthetic enzymes. There are several possibilities that may explain this discrepancy: (1) All the enzyme activities have been measured in microsomes derived from whole lung. Therefore, phospholipid synthesizing activities do not only represent surfactant synthesis. (2) The enzyme affinity for substrates may change during development. This possibility has recently been excluded [14]. (3) Acidic phospholipid content may be regulated by breakdown rather than synthesis rate. Again, present evidence does not support this alternative [20]. (4) Substrate concentrations at the biosyn-

Table II. Percentage of phospholipid composition of alveolar lavage fluid 2 days after myoinositol treatment

	Control	Myo-inositol treatment
Phosphatidylcholine	77.0	79.0
DPC	46.5	47.2
Phosphatidylinositol	5.8	9.9
Phosphatidylglycerol	6.1	0.3
Phosphatidylethanolamine	5.2	5.8
bis-(Monoacylglycerol)phosphate	1.4	0.8
Sphingomyelin	3.3	3.1
Phosphatidylserine	1.2	1.1

Myo-inositol treatment was started by continuous ear vein infusion of 12% of the sugar at a rate of 20 ml/kg/h for 2 h. Thereafter, the drinking water of the animals was supplemented with 8% of myoinositol, in addition to free access of standard rabbit chow.

thetic surface may determine the rates of the enzymatic reactions. The possible substrates are glycerol-free phosphate, myo-inositol, and CMP.

The last alternative was further studied. Glycerophosphate phosphatidyltransferase and CDP-diglyceride: inositol transferase activities were assayed simultaneously using microsomes from adults. *sn*-Glycerol-3-phosphate was present in excess (0.45 mM), whereas CDP-diglyceride was rate limiting (0.03 mM). Increase in myo-inositol concentration increased PI synthesis and correspondingly decreased glycerophosphate phosphatidyltransferase reaction, until the latter activity decreased to about 5% of the maximum. This competition for the common substrate was not detected when excess of CDP-diglyceride was available. Addition of excess CMP failed to increase PG synthesis, suggesting that the reverse reaction catalyzed by CDP: diglyceride-inositol transferase is functionally not important in the present conditions (pH 7.5).

To further study the role of myoinositol in regulation of acidic surfactant phospholipids, large amounts of myoinositol were infused into adult rabbits in order to increase cell concentrations to more than tenfold. After 2 days the lung fractions were isolated and analyzed for the phospholipid composition. Only surfactant fractions demonstrated a clear-cut change. As shown in table II, during myoinositol treatment PG practically disappeared, whereas PI increased correspondingly, and DPC remained unaffected.

These results indicate that CDP-diglyceride is rate limiting in the formation of the acidic surfactant phospholipids. The question still remains whether myoinositol has any physiological importance in controlling surfactant PG and PI during perinatal development. Preliminary evidence indicates that fetal serum levels of myoinositol exceed those in adults up to more than tenfold. Furthermore, myoinositol drops [5b] during the time period when PG appears and increases in quantity. Therefore, it is possible that this sugar is the most important factor that determines the quality of the acidic phospholipids associated with the surfactant complex.

Besides the dietary intake, the cellular levels of myoinositol may be regulated by biosynthesis, metabolism, and/or membrane transport. These aspects importantly remain to be studied.

Summary

Fetal accumulation of disaturated phosphatidylcholine into lung parenchyma preceeds its secretion into the future airways. In the human fetus, secretion of disaturated phosphatidylcholine increases from approximately the 28th week to term. Phosphatidylinositol increases parallel to disaturated phosphatidylcholine up to 36 weeks when phosphatidylglycerol appears. Phosphatidylglycerol improves the surfactant function and its presence indicates mature lungs.

The following sequence for activation of surfactant phospholipid synthesis in the fetus is proposed: the rate of acidic surfactant phospholipid synthesis is controlled by the availability of CDP-diglyceride. Increase in phosphatidate cytidyltransferase in immature fetus increases phosphatidylinositol. This phospholipid turns on phosphatidylcholine synthesis by direct activation of CDP-choline phosphatidyltransferase. Which of the acidic phospholipids, phosphatidylinositol or phosphatidylglycerol are synthesized, seems to be mainly controlled by myoinositol concentration at the biosynthetic surface.

The finding of *Gluck* and associates that the functional maturity of the fetus may be assessed by measuring the L/S ratio is of major importance in RDS prevention. Measurement of the lung profile, L/S ratio, percent disaturated phosphatidylcholine, percent phosphatidylglycerol, and percent phosphatidylinositol eliminates iatrogenic RDS and improves the management of the high-risk pregnancy.

References

1   Abe, M.; Akino, T., and Ohno, K.: On the metabolism of lecithin in lung and liver of fetal rabbits. Tohoku J. exp. Med. *109*: 163–172 (1973).

2   Adams, F.H. and Fujiwara, T.: Surfactant in fetal lamb tracheal fluid. J. Pediat. *63*: 537–542 (1963).

3   Avery, M.E. and Mead, J.: Surface properties in relation to atelectasis and hyaline membrane disease. Am. J. Dis. Child. *97*: 517 (1959).

4   Bangham, A.D.; Morley, C.J., and Phillips, M.D.: The physical properties of an effective lung surfactant. Biochim. biophys. Acta *593*: 552–556 (1979).

5a   Batenburg, J.J. and Van Golde, L.M.G.: Formation of pulmonary surfactant in whole lung and in isolated type II alveolar cells. Rev. perinat. Med. *3*: 73–114 (1979).

5b   Burton, L.E. and Wells, W.W.: Studies on the developmental patterns of the enzymes converting glucose-6-phosphate into myo-inositol in the rat. Devl Biol. *37*: 35–42 (1974).

6a   Clements, J.A.; Brown, E.S., and Johnson, R.P.: Pulmonary surface tension and the mucus lining of the lungs: some theoretical considerations. J. appl. Physiol. *12*: 262–268 (1958).

6b   Clements, J.A.; Nellenbogen, J., and Trahan, H.J.: Pulmonary surfactant and evolution of the lungs. Science *169*: 603–604 (1970).

7   Cunningham, M.D.; Desai, N.S.; Thompson, S.A., *et al.*: Amniotic fluid phosphatidyl-glycerol in diabetic pregnancies. Am. J. Obstet. Gynec. *131*: 719–724 (1978).

8   Farrel, P.M. and Morgan, T.E.: Lecithin biosynthesis in the developing lung; in Hodson, Development of the lung, pp. 309–347 (Marcel Dekker, New York 1977).

9   Feldman, D.A.; Kovac, C.R.; Dranginis, P.A., and Weinhold, P.A.: The role of phosphatidylglycerol in the activation of CTP: phosphocholine cytidyltransferase from rat lung. J.biol. Chem. *253*: 4980–4986 (1978).

10   Gluck, L.; Kulovich, M.V.; Borer, R.C., *et al.*: Diagnosis of the respiratory distress syndrome by amniotcentesis. Am. J. Obstet. Gynec. *109*: 440–445 (1971).

11a Gluck, L.; Kulovich, M.V.; Borer, R.C., *et al.*: The interpretation and significance of the lecithin/sphingomyelin ratio in amniotic fluid. Am. J. Obstet. Gynec. *120*: 142–155 (1974).

11b Gluck, L.; Sribney, M., and Kulovich, M.: The biochemical development of surface activity in mammalian lung. LL. The biosynthesis of phospholipids in the lung of the developing rabbit fetus and newborn. Pediat. Res. *1*: 247–265 (1967).

12   Hallman, M.; Feldman, B.H.; Kirkpatrick, E., and Gluck, L.: Absence of phosphatidyl-glycerol (PG) in respiratory distress syndrome in the newborn. Pediat. Res. *11*: 714–720 (1977).

13   Hallman, M. and Gluck, L.: Phosphatidylglycerol in lung surfactant. III. Possible modifier of surfactant function. J. Lipid Res. *17*: 257–262 (1976).

14   Hallman, M. and Gluck, L.: Formation of acidic surfactant phospholipids in rabbit lung during perinatal development. Pediat. Res. (in press 1980).

15   Hallman, M.; Kulovich, M.V.; Kirkpatrick, E., *et al.*: Phosphatidylinositol (PI) and phosphatidylglycerol (PG) in amniotic fluid. Indices of lung maturity. Am. J. Obstet. Gynec. *125*: 613 (1976).

16   Hallman, M. and Raivio, K.: Formation of disaturated lecithin through the lysolecithin pathway in the lung of the developing rabbit. Biol. Neonate *27*: 329–338 (1975).

17   Hallman, M. and Teramo, K.: Amniotic fluid phospholipid profile as a predictor of fetal maturity in diabetic pregnancies. Obstet. Gynec. *54:* 703–707 (1979).

18   Hamosh, M.; Shechter, Y., and Hamosh, P.: Metabolic activity of developing rabbit lung. Pediat. Res. *12*: 95–100 (1978).

19   Ikegami, M.; Silverman, J., and Adams, F.H.: Restoration of lung pressure-volume characteristics with various phospholipids. Pediat. Res. *13*: 777–780 (1979).

20 Jobe, A.; Kirkpatrick, E., and Gluck, L.: Labeling of phospholipids in the surfactant and subcellular fractions of rabbit lung. J. biol. Chem. *253*: 3810–3816 (1978).

21 Kikkawa, Y.; Motoyama, E.K., and Cook, C.D.: The ultrastructure of the lungs of lambs. The relation of osmiophilic inclusions and alveolar lining layer to fetal maturation and experimentally produced respiratory distress. Am. J. Path. *47*: 877–903 (1965).

22 Kulovich, M. and Gluck, L.: The lung profile II. Complicated pregnancy. Am. J. Obstet. Gynec. *135*: 64–70 (1979).

23 Kulovich, M.V.; Hallman, M., and Gluck, L.: The lung profile. I. Normal pregnancy. Am. J. Obstet. Gynec. *135*: 57–63 (1979).

24 Obladen, M.; Brendlein, F., and Krempien, B.: Surfactant substitution. Eur. J. Pediate. *131*: 219– 228 (1979).

25 Oldenburg, M. and Van Golde, L.M.G.: Activity of choline phosphotransferase, lysolecithin: lysolecithin acyltransferase and lysolecithin acyltransferase in the developing mouse lung. Biochim. biophys. Acta *444*: 433–442 (1976).

26 Pattle, R.E.: Properties, function and origin of the alveolar lining layer. Proc. R. Soc., Lond., Ser. B. *148*: 217–240 (1958).

27 Platzker, A.C.G.; Kitterman, J.A.; Mescher, E.H., et al: Surfactant in the lung and tracheal fluid of the fetal lambs and acceleration of its appearance by dexamethasone. Pediatrics, Springfield *56*: 554–561 (1975).

28 Schultz, F.M.; Jimenez, J.M.; MacDonald, P.C., and Johnston, J.M.: Fetal lung maturation. I. Phosphatidic acid phosphohydrolase in rabbit lung. Gynec. Invest. *5*: 222–229 (1974).

29 Stern, W.; Kovac, C., and Weinhold, P.A.: Activity and properties of CTP: cholinephosphate cytidyltransferase in adult and fetal rat lung. Biochim. biophys. Acta *441*: 280–293 (1976).

30 Young, A.; Casola, P.G.; Wang, C., et al.: Pulmonary phosphatidic acid phosphatase. A comparative study of the aqueously dispersed phosphatidate dependent and membrane-bound phosphatidate dependent phosphatidic acid phosphatase activities of rat lung. Biochim. biophys. Acta *574*: 226–239 (1979).

M. Hallman, MD, University of California, San Diego, La Jolla, CA 92093 (USA)

Discussion

   Hallman on a question of *van Golde,* did not have an explanation about the existing species differences in lipid metabolism. It might be relevant how the enzyme activities have been analysed. *Hallman* agreed with *van Golde* that *in vitro* experiments on PI-PG competition experiments should be carried out on type II cells. However, the fact that the acidic phospholipids only change in surfactant strongly imply that myoinositol-sensitive synthesis takes place mainly, if not exclusively, in type II cells *in vivo.* In answering a further question of *van Golde, Hallman* stated that the inositol phosphotransferase reaction is reversible but the back-reaction takes place effectively at low pH in the presence of high CMP and low CTP. Choline phosphotransferase is reversible, too, and high CMP may consequently paradoxically inhibit PC synthesis, too. It is unlikely that CMP regulates surfactant phospholipid synthesis. *Hallman* agreed with the comment of *Clements* that there are species differences in metabolism of PG and PI.

   Mietens wondered if there were an influence of acidosis on the rate of synthesis of surfactant in animals. *Hallman,* referring to the studies of *Merrit,* was not quite convinced how much pH affects these mechanisms; processes other than related to surfactant may be more sensitive to acidosis. *Clements* asks *Hallman* and *van Golde* about the significance of the methylation pathway in PC synthesis, but this pathway appears to *van Golde* to be of minor importance although the activity of this pathway can be demonstrated in the lung. The data on the isolated type II cells show a great preponderance of the choline pathway, so it seems that the methylation pathway is, at least in the adult animal, not very important.

Prog. Resp. Res., vol. 15, pp. 41–48 (Karger, Basel 1981)

Mechanical Development in Fetal Lungs

R. Rüfer

Institute for Pharmacology and Toxicology, Faculty for Clinical Medicine Mannheim of the University of Heidelberg, Mannheim

Introduction

The change from the solid to the canalicular stage of the lungs in man was estimated to occur in the 16th week of gestation [4]. At this time the developing capillaries relatively suddenly begin to interrupt the cubic epithelium of the prospective airways. The epithelium itself becomes more and more flat and the morphological appearance of the lungs becomes canalicular. In different species this point of transition is entirely different. In terms of the total time of gestation in man the lungs become 'opened' after about 40% of this period while in rabbits for example the canalicular stage starts not before 75% of the gestation.

This paper deals with the mechanical development of fetal mini-pig lungs. The period of gestation of this species continues for 110–112 days. We roughly estimated the canalicular stage to appear after 82–84 days which corresponds to 75% of the total gestation. From the morphological point of view this is the earliest moment at which we can expect the possibility of an inflation of these lungs. However, in order to compare this virtual date with the real point of lung expansion we studied the mechanical development of fetal lungs using pressure-volume techniques.

Methods

Recording of Pressure-Volume Diagrams

Fetal mini-pigs of different gestational ages (95, 100 and 110 days) were obtained by cesarean section. The lungs were removed and static pressure-volume (p-v) curves of the isolated lungs were recorded in succession using both air and liquid for inflation. Therefore, in fetal lungs we did not see differences depending on the consecutive sequence of the procedure: p-v curves using air for inflation recorded before liquid filling showed no substantial differences from

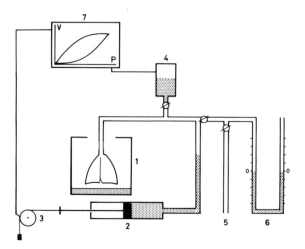

Fig. 1. Experimental set-up used for the recording of static pressure-volume diagrams. 1 = Lung chamber with the lungs suspended in humid air respectively with the lungs immersed in saline solution; 2 = infusion pump; 3 = volume gauge; 4 = pressure gauge; 5,6 = device for the preceding opening inflation; 7 = X-Y recorder.

curves recorded after filling the lungs with liquid. However, since the fetal lungs *a priori* are atelectatic, respectively, filled with amniotic fluid, we usually started with the liquid p-v diagram. The actual recording of the ▶-v diagram was preceded by an opening inflation with liquid, respectively, air.

p-v Diagrams Using Air (fig. 1). Starting at atmospheric pressure the lungs were continously inflated with air up to an end-inspiratory pressure of 25 cm H_2O which in adult lungs corresponds with the total lung capacity for air. Vice versa the volume was removed by the same modus down to the starting pressure. The volume was changed with 120 ml/h which was proved before to be quasi static.

p-v Diagrams Using Liquid. The system shown in figure 1 was completely filled with saline solution. In order to compensate hydrostatic pressure differences the lungs were immersed in the liquid. Because of the eliminated surface forces the lungs were expanded with saline solution up to 8 cm H_2O only, which corresponds to the total capacity for liquid of adult lungs [3]. In order to maintain quasi-static conditions, the volume change was reduced to 6 ml/h because of the increased resistances during liquid filling.

Evaluation of p-v Diagrams

Compliances. We estimated the compliances in both air-filled and liquid-filled lungs according to the relationship:

$$C = \frac{\Delta v}{\Delta p} \left[\frac{ml}{cm\ H_2O} \right],$$

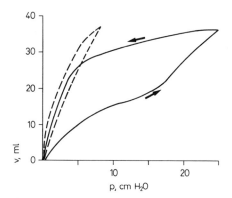

Fig. 2. Static pressure-volume curves of the lungs of a newborn mini-pig. ——— = Air used for inflation; – – – – = saline solution used for inflation.

(where Δv = volume change; Δp = pressure change, in liquid-filled lungs = 0–8 cm H_2O, in air-filled lungs = 0–25 cm H_2O). According to *Radford* [3], we corrected for the different lung sizes of the mini-pigs by expressing the compliance per unit of lung dry weight which was measured in each individual lung. The parameter $\dfrac{\Delta v}{\Delta p \cdot g}$ was called weight-specific compliance (C_{spec}).

Estimation of Lung Retraction. The resistances against lung compliance reflect the retractive forces of the lungs directly. Therefore, the reciprocal value of the specific compliance measured under static conditions is the static specific lung retraction $R = \dfrac{1}{C_{spec}}$.

According to *van Neergaard* [2], the overall retraction (R_{tot}) measured by filling the lungs with air minus the retraction due to tissue elasticity (R_{tis}) is the retraction caused by surface forces (R_{surf}) · $R_{surf} = R_{tot} - R_{tis}$.

Results

Figure 2 shows the p-v diagrams of the lungs of a newborn mini-pig after the regular gestation period of 112 days. Compared to the inflation with liquid the inflation with air needs approximately 3 times higher pressures for the expansion of the lungs to the same total volume.

Expressed in terms of lung *compliances* in this individual example shown in figure 2, there are remarkable differences: the compliance for air is estimated to be 1.5 ml/cm H_2O while the compliance for liquid in the same lungs is 5.0 ml/cm H_2O, that is, a ratio of 1 : 3.4.

Comparing different lungs of the same gestation period in regard of their *specific compliances,* we found in the end of the regular gestation period (110 days) a ratio of 1:5.0 air versus liquid (fig. 3).

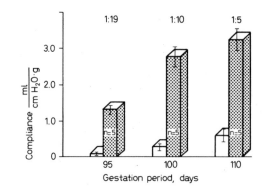

Fig. 3. Compliances per gram dry weight (specific compliances) of fetal mini-pig lungs at different gestation periods. White columns = Compliances for air; dotted columns = compliances for saline solution. On the top of the columns the ratio between both compliances is given.

The more immature the lungs, the more this ratio increases as shown in figure 3. At 85% of the total gestation period (day 95) the lungs are almost noninflatable for air. However, despite the biological immaturity there is a volume capacity for liquid 19 times higher than for air.

As indicated in table I, both of the specific lung compliances increase during the prenatal development of the lungs, i.e. the compliance for air and for liquid. What differs is the extent of this increase. Between days 95 and 110 of gestation the specific compliance for liquid increases by factor 2.3, while the specific compliance for air increases by factor 8.4 during the same period. During the first 7 days of the postpartal period comparable changes of the lung compliances could not be observed (table I).

In another approach, the analysis in terms of *lung retraction* given in table II directly shows the influence of the maturity on the mechanical behavior of the lungs. The two components of overall lung retraction decrease during prenatal life. While the retraction due to tissue elasticity decreases by factor 2.3 (from 0.68 to 0.30) during the last 15 days of the gestation period the retraction due to surface forces decreases by factor 9.9 (from 11.82 to 1.19) during the same period. Expressing the retraction due to surface forces in percent of the total retraction, we observed a remarkable decrease between days 95 and 110 of the gestation. In the beginning of our observation period at day 95 of the gestation, about 95% of the total retraction is due to surface forces while only the remaining 5% are due to tissue elasticity. At the end of the gestation one-fifth of the total retractive

Table I. Specific compliances (C_{spec}) for air and liquid in fetal mini-pig lungs (A) compared with the C_{spec} of newborn (B): mean ± standard error of the mean

Gestation period/postpartal period, days	C_{spec} air $\dfrac{ml}{cm\ H_2O \cdot g}$	C_{spec} liquid $\dfrac{ml}{cm\ H_2O \cdot g}$
A		
95	0.08 ± 0.02	1.46 ± 0.12
100	0.24 ± 0.07	2.64 ± 0.29
110	0.67 ± 0.15	3.28 ± 0.24
B		
7	0.69 ± 0.18	3.60 ± 0.54

Table II. Retraction ($\frac{1}{C_{spec}}$) of fetal mini-pig lungs (A) compared with the retraction of newborn (B).

Gestation period/ postpartal period, days	R_{tot} $\dfrac{1}{ml/cm\ H_2O \cdot g}$	R_{tis} $\dfrac{1}{ml/cm\ H_2O \cdot g}$	R_{surf} $\dfrac{1}{ml/cm\ H_2O \cdot g}$	R_{surf}/R_{tot}, %
A				
95	12.50	0.68	11.82	94.6
100	4.17	0.38	3.79	90.9
110	1.49	0.30	1.19	79.9
B				
7	1.45	0.28	1.17	80.7

Calculation on the basis of the mean values of table I. R_{tot} = Total retraction; R_{tis} = retraction due to tissue elasticity; R_{surf} = retraction due to surface forces.

forces are caused by tissue elasticity, that means 80% of the total retraction is caused by surface forces. However, as shown in table II, we could not observe a further decrease during the first 7 days after birth.

Discussion

It has to be mentioned that due to the immaturity and fragility of the tissue it is difficult to obtain p-v curves of fetal lungs. Furthermore, we have to realize the methodological problems in the omparison of compliances or retractions of lungs subsequently being filled with liquid and air or vice versa.

The condition to be static in both techniques was taken into account by the variation of the infusion velocity. Although our system was checked to be quasi-static in both liquid and air-filling, we have observed small hysteresis effects even in liquid-filled lungs (fig. 2). These effects may be explained by alveolocapillary liquid transfer as well as by temperature effects or by leakages. *Agostoni et al.* [1] have explained similar hysteresis effects in static pressure-volume diagrams of mature fetal lungs with plastic qualities of the fetal lung tissue.

Comparing the end-inspiratory volumes of lungs filled with liquid and with air, it would be desirable to refer to equal pressures. Due to biological reasons this is impossible as far as the lungs have to be inflated to their total capacities in both conditions. Therefore, the total capacity is defined as a condition in which a further increase of the intrapulmonary pressure is not answered by a further uptake of volume. Using liquid for inflation this total capacity is equal to a pressure of about 8 cm H_2O; using air for inflation the total capacity is reached at about 25 cm H_2O. Therefore, using the specific compliance, respectively, their reciprocal, we have referred the volume per cm H_2O and for correction of differences in the water content we have referred per gram dry weight.

Our results describe as an approach the volume history of lungs being extremely expanded to their total capacity. But even in more physiological ranges of lung expansion there are remarkable differences between liquid and air-filled lungs as the p-v curves show (fig. 2).

The data describe the mechanical development of fetal mini-pig lungs during the last 15% of the gestation period. Although in this species the beginning of the canalicular stage was estimated to start with days 82–84 of the gestation period, the lungs are noninflatable with air up to day 95 or later. However, there already is detectable a volume capacity for liquid. Consequently, we have to differentiate between these two expansions: expansion for air and expansion for liquid.

Even at the end of the normal gestation period a square unit of lung tissue has a capacity for liquid 5 times higher than for air. This result is in agreement with *van Neergaard's* [2] observation obtained in adult pig lungs and is explained with the influence of the surface tension on the lung retraction. The lower the content of surface-active substances in immature lungs, the higher is the retraction due to surface forces. The comparison

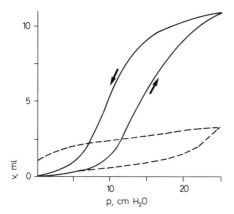

Fig. 4. Static pressure-volume curves of the lungs of two fetal mini-pigs at day 95 of the gestation period. Identical lung dry weights and phospholipid contents. – – – = After 20 min artificial ventilation of the fetus with air; ——— = After 20 min artificial ventilation of the fetus with Fluorocarbon FX 80.

of earlier results on the content of phospholipids in fetal lungs [5] with the results concerning the mechanical resistances against inflation shows a direct correlation between lung inflation and phospholipid content.

The more subtle analysis of the relationship between the two components of the overall retraction in our data shows that both components decrease during the observation period. What differs is the degree of this decrease. The retraction due to tissue elasticity decreases by factor 2.3, while the retractive forces due to surface forces decrease approximately 10 times during the last 15% of gestation.

Discussing therapeutical aspects in the treatment of respiratory distress, it could be useful to consider this observation. We should substitute the two functions of the original alveolar surfactant: the first function of lowering the surface tension in general as well as the second function of the stabilization of alveoli having different diameters.

The influence of the surface tension on the compliance of immature lungs can be demonstrated by experiments published earlier [6]. Figure 4 shows the p-v curves of the lungs of 2 immature fetuses with the same gestational age. After artificial ventilation of the fetus with Fluorocarbon FX 80, the low surface tension of 15 mN/cm of this compound enables the inflation of the lungs. Although in terms of gestational age these lungs are immature, their compliance is normalized to the values of mature newborn.

Summary

The development of the mechanics of the lungs was studied in fetal mini-pigs of different gestation periods. Using static pressure-volume techniques the specific compliances were estimated for both liquid and air. During the last 15% of gestation, between day 95 and 110 of the total period, the specific compliance for liquid increased by factor 2.3 while the specific compliance for air increased 8.4-fold. In the beginning of this period, 95% of the total lung retraction was found to be due to surface forces; the remaining 5% was calculated to be due to tissue elasticity. At the end of the gestation period 80% of the total retraction was caused by surface forces and 20% by tissue elasticity.

References

1   Agostoni, E.; Taglietti, A.; Agostoni, A.F., and Setrikar, I.: Mechanical aspects of the first breath. J. appl. Physiol. *13:* 344–348 (1958).
2   Neergaard, K. van : Neue Auffassungen über einen Grundbegriff der Atemmechanik. Die Retraktionskraft der Lunge, abhängig von der Oberflächenspannung in den Alveolen. Z. ges. exp. Med. *66:* 373–394 (1929).
3   Radford, E.P., jr.: Static mechanical properties of mammalian lungs; in Fenn and Rahn, Handbook of physiology; pp. 429–449 (American Physiological Society, Washington 1964).
4   Reid, L.: The embryology of the lung; in de Reuck and Porter, Development of the lung, pp. 109–124 (Churchill, London 1967).
5 · Rüfer, R.: The influence of surface active substances on alveolar mechanics in the respiratory distress syndrome. Respiration *25:* 441–457 (1968).
6   Rüfer, R. and Spitzer, H.L.: Liquid ventilation in the respiratory distress syndrome. Chest *66:* suppl. pp. 29–30 (1974).

Dr. R. Rüfer, Institut für Pharmakologie und Toxikologie der klinischen Fakultät Mannheim der Universität Heidelberg, Maybachstrasse 14-16, D-6800 Mannheim 1 (FRG)

Discussion

Some questions by *Mitzner* concerned the pressure used in the experiments to inflate the lungs. It was, *Rüfer* answered, 30 cm H_2O containing air and 8–10 cm H_2O using saline, higher pressures lead to rupture of alveoli. *Enhorning* commented on the fact that pulmonary surfactant in the neonate has two functions. It helps the initial aeration of the lung and secondly, and perhaps more importantly, it stabilizes the lung. He wondered if fluorcarbon functioned as a detergent rather than a stabilizer. *Rüfer* agreed that fluorcarbon does not act like a film, and *Clements* also agreed that one should probably distinguish between agents which can lower surface tension to very low values and detergents. *Rüfer* answering a question by *Robertson* said that the experiments seemed to show no severe damage of the alveolar epithelium. *Tiebes* asked if FC 75 induced a release of surface-active material into the alveoli, but *Rüfer* could find only traces of surfactant material, so there was practically no loss of surfactant.

Prog. Resp. Res., vol. 15, pp. 49–56 (Karger, Basel 1981)

Surface Properties of Pulmonary Surfactant – Evolutionary and Physiological Aspects

R. Reifenrath, A. Vatter and C. Lin

Universitätsklinik 'Bergmannsheil' Bochum; Webb-Waring Lung Institute, Denver, Colo.; Departements of Pediatrics and Gynecology and Obstetrics, Albert Einstein-College of Medicine, and Bronx Municipal Hospital Center, New York, N.Y.

During surface tension measurements of mixed lecithin-cholesterol films, extreme differences in the kinetics of film formation became evident. While the film formation of certain lecithin-cholesterol ratios required many hours, the final film of others was formed within minutes. Because of this empirical observation we introduced a kinetic parameter in a subsequent study. The trial of this study was to compare different surface tension parameters of human amniotic fluid with gestational age.

Figure 1 illustrates the term 'film formation'. Simulating the creation of the bronchio-alveolar air/liquid surface after birth, we created an air-liquid surface with the amniotic fluid as subphase. After onset of surface oscillation, minimum surface tension decreases with time until a final stable value is reached. The delta-gamma, i.e. the difference between surface tension before onset of oscillation and final minimum surface tension, divided by t, i.e. the time between onset of oscillation and the observation of final minimum surface tension, was chosen as parameter reflecting the kinetics of film formation. Figure 2 reveals the relationship between this parameter and gestational age.

The experiments were not designed to give a high resolution with respect to this parameter. The minimum surface tension, for example, was recorded each 10th min during oscillation, sometimes 5 min after the oscillation had begun. The rate of decreasing minimum surface tension given in figure 2 for the sample representing the 41st week was obtained 5 min after onset of oscillation, but actually the minimum surface tension was reached after 30 sec. Thus, the rate is about 10 times greater than the 4.02 mN/m/min given in figure 2. This means, film formation was completed

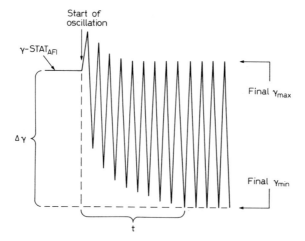

Fig. 1. Schematic representation of the change in minimum surface tension during oscillation. $\gamma\text{-STAT}_{AFI}$ = static surface tension 2 min after formation of the interface; γ_{min} = minimum surface tension; $\Delta\gamma$ = difference between $\gamma\text{-STAT}_{AFI}$ and final γ_{min}: t = time interval between start of oscillation and the first observation of final γ_{min}.

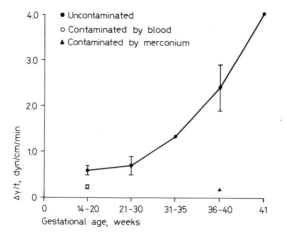

Fig. 2. Relation between rate of decreasing γ_{min} during oscillation at 25% area variation (γ/t) and gestational age. Mean values $\pm$ SE. For explanation of symbols see figure 1.

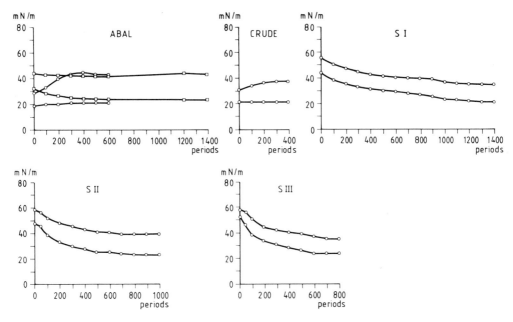

Fig. 3. Kinetics of film formation represented by the number of periods necessary to obtain a stable area/surface-tension diagram. Surfactant preparations (rabbit) obtained by the courtesy of Dr. *O. Reiss,* Denver. ABAL = acellular bronchio-alveolar lavage; Crude = crude extract; SI–SIII = preparations after zonal centrifugation. Phospholipid concentration of subphase (phosphate buffer) 4 μg/ml (□) and 53 μg/ml (○) for ABAL and 20 μg/ml for Crude, SI, SII and SIII. Temperature 37°C. 25% area variation, 3 sec/period.

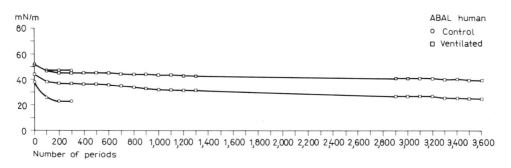

Fig. 4. Kinetic of film formation represented by the number of periods necessary to obtain a stable area/surface-tension diagram. Surfactant preparations obtained by the courtesy of Dr. *Reiss.* ABAL = acellular bronchio-alveolar lavage. Phospholipid concentration of subphase (phosphate buffer) 20 μg/ml. Temperature 37°C. 25% area variation, 3 sec/period.

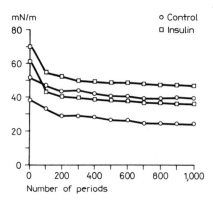

Fig. 5. Kinetics of film formation represented by the number of periods necessary to obtain a stable area/surface-tension diagram. Surfactant preparations (sheep fetus) obtained by the courtesy of Dr. *Reiss.* Phospholipid concentration of subphase (phosphate buffer) 20 µg/ml. For physical conditions see figure 3.

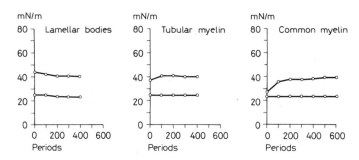

Fig. 6. Kinetics of film formation represented by the number of periods necessary to obtain a stable area/surface-tension diagram. Surfactant preparations obtained by the courtesy of Dr. *R. Sanders,* Denver. Phospholipid concentration of subphase (phosphate buffer) 20 µg/ml. For physical conditions see figure 3.

within 30 sec at 40 weeks and 30 min at 30 weeks of gestation, i.e., $\delta\gamma/t$ increases by a factor of 60 between the 30th and 40th week of gestation! It seems to be likely that this parameter is very important during neonatal adaptation to air breathing. In order to underline this statement, two comments should be added. Firstly; different other static and dynamic surface tension parameters did not depend upon gestational age. Secondly,

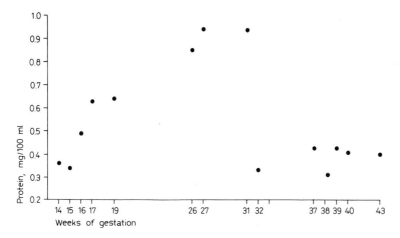

Fig. 7. Protein concentration of human amniotic fluid at different gestational weeks.

in earlier studies we found an agreement of surface tension properties between pulmonary surfactant and mixed lecithin-cholesterol films in certain weight ratios. This agreement still holds for the static and dynamic surface tension – but not for the kinetics of film formation. Natural surfactant films, including films formed by tubular myelin, common myelin and lamellar bodies, are formed within seconds (comparable to amniotic fluid at term), the mixed films within minutes (comparable to amniotic fluid at the 30th week of gestation).

The kinetics of film formation certainly is not simply a function of phospholipid concentration. The results represented by figures 3–6 are based on the same concentration of phospholipids. However, the kinetics of film formation differs considerably.

In the case of amniotic fluid there is also no correlation between kinetics of film formation and protein concentration (fig. 7). Beside a changing lecithin-cholesterol ratio during gestation [unpublished observation], no explanation has been obtained indicating the origin of the factor(s) influencing the kinetics of film formation.

In contrast to amniotic fluid, surface-active materials derived from the lung directly revealed one other parameter changing with gestational age. Human fetal pulmonary fluid obtained by alveolar micropuncture revealed a significant change in minimum surface tension measured at different amplitudes of area oscillation only in case of immaturity (fig. 8).

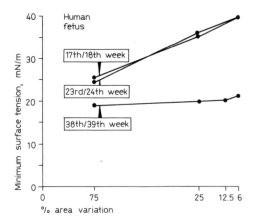

Fig. 8. Minimum surface tension during the final area/surface-tension diagram (ASD). Samples delivered onto the surface. (Sample obtained by micropuncture, final ASD, 37°C, 0.9% NaCl solution.)

This ontogenetic aspect of the lungs surface activity might be correlated with results of a phylogenetic study, based upon the hypothesis that a lung composed of alveoli would never have been evolved with the lung surfactant absent. Figure 9 represents minimum surface tensions during final area/surface-tension diagrams at different area variations of lung lavage obtained from lung fish, salamander, snake, frog, dog and human. While this parameter in figure 8 was correlated with gestational age, figure 9 reveals a correlation with the development of the alveolar lung type. The lung fish has no alveoli, the salamander and the snake have increasing folds along their lung surface, the snake develops first alveolar-like structures at the blind end of lung. In contrast to the species mentioned, the frog has the first layers of alveoli covering the entire lung surface.

Summarizing the data presented, in the first place we emphasize the importance of surface tension measurements as an excellent tool in analyzing the lungs surfactant system. According to a large body of experimental findings, this valuable tool is critically dependent on a few boundary conditions: the air at the surface must be saturated by water vapor; the film formation has to occur from the hypophase if kinetic studies are desired; area variations significantly smaller than 75% are obligatory in order not to overlook different properties of different materials; the final, stable film has to be reached form different reasons.

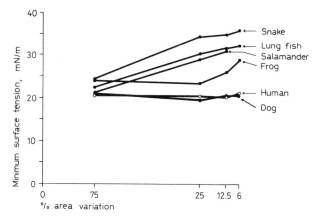

Fig. 9. Minimum surface tension during the final area/surface-tension diagram (ASD). Human lung lavage (final ASD, 37°C, Ringer's solution) obtained by the courtesy of Dr. *Reiss.* Values for dog taken from *Slama et al.:* Resp. Physiol. *19:* 233–243 (1973).

If care is taken in observing these rules, the kinetics of film formation and the minimum surface tension at different amplitudes of area oscillation may serve as reliable measures determining the role of surfactant under normal conditions and in certain disorders of lung function. The applicability of the method is not limited by a surface tension measuring device complicated like Slama's bubble apparatus. The rate of decreasing static surface tension after a standardized and rapid formation of the surface may serve as a kinetic parameter as well as the rate of decreasing minimum surface tension during oscillation.

Summary

Surface tension measurements on human amniotic fluid, human fetal pulmonary fluid, lung lavage of lung fish, salamander, snake, frog, dog and human revealed a change of surfactant properties related to the phylogenesis and ontogenesis. The only surface tension parameter changing with phylognenesis or ontogenesis was kinetic of film formation and/or minimum surface tension. These two parameters can be obtained properly only by consideration of different experimental conditions.

R. Reifenrath, MD, Berlinerstrasse 42, D–3104 Unterlüss (FRG)

Discussion

In reply to a question from *Morley, Reifenrath* confirmed that he had used exclusively dynamic studies with 3–sec periods and he had seldom observed a minimum of surface tension below 20 dyn/cm. He answered an additional question by *Tierney,* with the information that the minimum surface tension is due to the kinetic differences which often differ before the final stable film has been reached. *Baum* also commented that temperature and equilibration times are very important in measuring surface tension. *Morley* disagreed with some remarks, because he had been able to achieve a surface tension of zero with natural surfactant compressed 20 times/min and 37°C.

Prog. Resp. Res., vol. 15, pp. 57–61 (Karger, Basel 1981)

Pulsating Bubble Technique for Evaluating Pulmonary Surfactant[1]

Goran Enhorning

Department of Obstetrics and Gynaecology, University of Toronto, Toronto, Ont.

Surface tension is manifested in many ways. Capillarity is one. If a glass capillary is lowered into a water solution, or suspension, the liquid will rise inside the capillary to a certain height, h (fig. 1a). In order to make the air-liquid interface move down again to the level of surrounding liquid, one would have to exert a certain pressure, P. The force required is the pressure, P, multiplied by the inner cross section of the capillary, πr^2 (fig. 1b), and that force is counteracted by surface tension, γ, acting along the inner circumference of the capillary, $2\pi r$. From the equation, $P\pi r^2 = \gamma 2\pi r$, we get the law of Laplace as it applies for a spherical surface: $P = 2\gamma/r$. If we wish to move the meniscus all the way down to the capillary tip, P would then have to be raised continuously to overcome the effect of gravity. Suddenly, however, the value of P would drop dramatically as a bubble is created. The radius of the bubble, R, is much larger than that of the capillary, r, hence the sudden drop in value of P (fig. 1c). At this point the system would be unstable, the bubble would continue to increase in size until it dislodged and water again entered the capillary. A stable system would be created though if the capillary tip were enclosed in a liquid-filled chamber (fig. 1d). Slow evacuation of that chamber would create a bubble, the size of which could be accurately controlled and measured through a microscope. Liquid could be made to move in and out of the chamber at a certain rate, thereby cyclically changing the bubble from maximal to minimal size in a precisely known fashion. If pressure is recorded in the liquid of the chamber, its absolute value will be equal to the pressure gradient across the bubble, since the latter communicates with

[1] Supported by Medical Research Council of Canada Project MT-4497

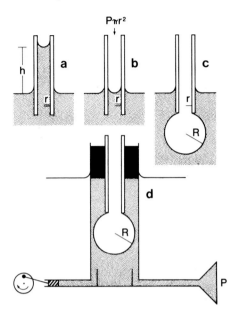

Fig. 1. Cross section of a glass capillary lowered into water solution or suspension. Liquid will rise to a certain height, h (a). The force $P\pi r^2$ is required to make the meniscus move down to the level of the surrounding liquid (b). If P is raised further, the meniscus will move down to the capillary tip, but then, as a bubble expands, the value of P drops instantaneously (c). The principle of the pulsating bubble technique is to record pressure across the surface of a bubble, the radius of which is measured through a microscope and oscillates between 0.55 and 0.4 mm (d). Knowing P and r, surface tension, γ, can be calculated from the law of Laplace, $P = 2 \gamma/r$.

ambient air. This is the principle of the pulsating bubble technique, using the law of Laplace to calculate the value of surface tension, γ, when the value of r is a known entity, and P is recorded [1–4]. The Surfactometer, the apparatus developed in my laboratory to measure surface tension according to this principle, has sample chambers with a capacity of no more than 20µl. The chambers are made of nylon tubing. To ensure correct temperature, usually 37°C, the chamber is lowered in thermostat-controlled water, which also serves the purpose of eliminating refraction and reflection in the cylindrical surface of the chamber. This makes it possible to clearly view the bubble through a microscope as it pulsates from maximal to minimal size (fig. 2). The pulsation rate is 20 rpm, i.e. it takes 3 sec for a full cycle. The radius is 0.55 and 0.4 mm at maximal and

Fig. 2. Air bubble photographed inside the chamber containing 20 µl of sample liquid to be examined. The bubble changes from maximal size (r = 0.55 mm) to minimal (r = 0.4 mm) in 1.5 sec.

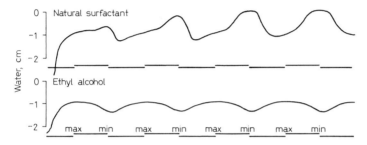

Fig. 3. Pressure tracings obtained when bubble is pulsating in surfactant *suspension* from rabbit lung wash (10 mg phospholipids/ml) and in ethyl alcohol. In the third cycle after creation of the bubble in natural surfactant, surface tension is zero and approximately 30 mN·$^{-1}$ at minimal and maximal bubble size, respectively. When the bubble is expanded in a *solution* of surface active material (alcohol), the tracing reflects a change in bubble radius, but not in the value of surface tension.

minimal bubble size, respectively. The surface area at minimal bubble size is 50% of what it is at maximal size. This is much less of a surface compression than is used in the Langmuir-Wilhelmy system and yet, when natural surfactant is tested with the pulsating bubble technique, a surface tension of zero is recorded at minimal bubble size, in fact even before surface area has reached its minimum.

The pulsating bubble technique offers several advantages. Apart from the small sample volume required, 20 µl, it clearly demonstrates the dynamics of natural surfactant, obtained from lung lavage of healthy rabbits. Almost instantaneously, natural surfactant will form a monomolecular film in the air-liquid interface, giving a maximal surface tension of around 30 mN·m^{-1} and a value of zero at minimal bubble size (fig. 3). An artificial surfactant preparation would have to show the same dynamics to be able to facilitate lung expansion in the neonate.

With special attachments, it is also possible with the pulsating bubble technique to introduce agents that will affect the function of the mono-molecular surface layer. The agents can be delivered to either the hypo-phase or the gas phase. An example of the latter type of interference is the introduction of solvent vapours into the bubble as it is pulsating. When vapours in a concentration of approximately 35% were allowed to enter the bubble during exactly ten cycles, it was found that chloroform and halothane had a dramatic effect. Surface tension changed almost instan-taneously to about 25 mN·m^{-1} and stayed at that value throughout the cycle. This clearly indicated an instability which, however, was not notice-able to the same extent with vapours of fluothane or acetone, and not at all with ether. Perhaps this partly explains the safety with which ether can be used as an anaesthetic and the potentially lethal risk of chloroform and halothane. At concentrations less than 20%, neither chloroform nor halo-thane had any effect. Hence, at concentrations used for anaesthesia there is no indication that any of the vapours tested woud have a detrimental effect on surfactant, at least not during the short period of testing, 30 sec.

Summary

The principle of this technique is to record the pressure gradient across the surface of an air bubble which is pulsating in 20µl of the liquid to be tested. The radius is then oscillating between 0.55 and 0.4 mm. Knowing pressure gradient and radius, surface tension can be calculated with the law of Laplace. The method is particularly suited for a study of surface kinetics. With pulmonary surfactant, from rabbit lung washings, a surface film is formed instantaneously and, within a few pulsations, surface tension is lowered to zero. When high concentrations of certain anaesthetic vapours (halothane, chloroform) are introduced into the pulsating bubble, surface properties of pulmonary surfactant are disrupted. The same concen-trations of other vapours (ether, fluothane, acetone) have minimal effect.

References

1   Adams, F.H. and Enhorning, G.: Surface properties of lung extracts. I. A dynamic alveolar model. Acta physiol. scand. *68:* 23–27 (1966).
2   Nozaki, M.: Pressure-volume relationship of a model alveolus. Tohoku J. exp. Med. *101:* 271–279 (1970).

3   Slama, H.; Schoedel, W. und Hansen, E.: Bestimmung der Oberflächeneigenschaften von Stoffen aus den Lungenalveolen mit einer Blasenmethode. Pflügers Arch. *322:* 355–363 (1971).

4   Enhorning, G.: Pulsating bubble technique for evaluating pulmonary surfactant. J. appl. physiol. *43:* 198–203 (1977).

G. Enhorning, MD, Department of Obstetrics and Gynaecology, University of Toronto, Toronto, Ontario (Canada)

Discussion

Dr. *Morley* asked about the role of halothane in causing severe changes in the physical properties of surfactant. *Enhorning* pointed out that on entering the bubble the vapours would first come in contact with the hydrophobic fatty acids of the phospholipid molecules in the monolayer. Halothane and chloroform are good solvents of lipids with saturated fatty acids. Ether, on the other hand, is a poor solvent of disaturated phospholipids, and had minimal effect on surfactant. This would imply that the main component in the surface monolayer was a disaturated phospholipid, probably DPPC.

Prog. Resp. Res., vol. 15, pp. 62–75 (Karger, Basel 1981)

Role of Surface Tension and Surfactant in the Transepithelial Movement of Fluid and in the Development of Pulmonary Edema[1]

Arthur C. Guyton and David S. Moffatt

Department of Physiology and Biophysics, University of Mississippi Medical Center, Jackson, Miss.

Several studies have shown the normal pulmonary membrane to be sufficiently discontinuous that rapid equilibration occurs between the fluids in the alveoli and the fluids in the interstitium. For instance, *Staub et al.* [1] demonstrated that radioactively labeled albumin, when instilled into the respiratory passages, enters the adjacent interstitium and becomes equilibrated between alveolar fluid and interstitial fluid within approximately 30 min. Recent measurements by *Parker et al.* [2] in our laboratory of the reflection coefficient of the alveolar membrane to albumin gave an average value of only 0.23, which fits with the findings of *Staub et al.*[1]. Therefore, if it is true that fluids within the alveoli exist normally in a state of equilibrium with the fluids of the pulmonary interstitium, then it also follows that the pressures of the fluids on the two sides of the membrane must also be equal. On first thought, one might doubt this because intuition tells us that the pressure of the fluids within the alveoli is equal to atmospheric pressure. But, as we shall see, the surface tension of the fluid in the alveoli makes this untrue, especially so where there are accumulations of fluid in alveolar corners, alveolar apices, crevices, and even in membrane pores.

Furthermore, the reduction of surface tension by surfactant is probably one of the most important factors in preventing the development of pulmonary edema. Therefore, let us discuss in the following sections of this paper some of the basic relationships between surface tension and fluid pressures in the alveoli, as well as the relationship of these to the development of pulmonary edema.

[1] Supported by grant-in-aid No. HLB 11678 from the USPHS.

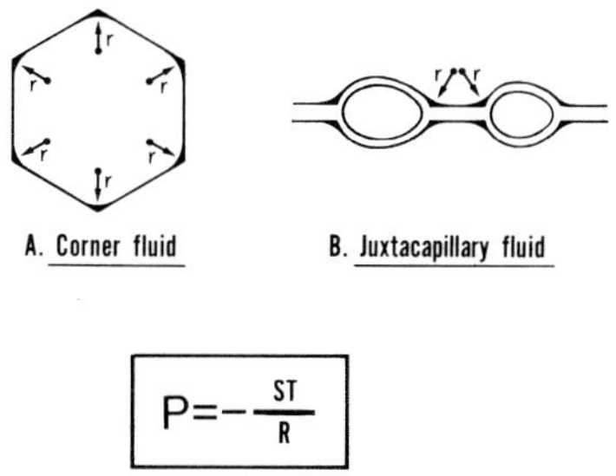

A. Corner fluid   B. Juxtacapillary fluid

$$P = - \frac{ST}{R}$$

Fig. 1. Accumulation of fluid in alveolar corners (A) and in the angles adjacent to capillary bulges in the alveolar wall (B). The air-fluid surface that is formed is cylindrical.

Effect of Surface Tension on Fluid Pressures under Curved Fluid Surfaces

Negative Pressures in Alveolar Corners and in the Angles along-side Capillary Bulges

Figure 1 (A) illustrates an alveolus with collections of fluid in the corners between adjacent walls. The apices that are formed where three walls meet will be discussed separately later. In corners, the fluid forms a cylindrical surface because the surface tension tends to pull the fluid toward the middle of the alveolus. That is, the surface is circular in one dimension but straight in the other dimension. When this is the case, the negative pressure beneath the surface of the fluid (with respect to alveolar air pressure) can be calculated using the following formula:

$$P = - \frac{ST}{R} \qquad (1)$$

in which P is the fluid pressure in dynes per square centimeter when the pressure is referred to atmospheric pressure, ST is surface tension in dynes per centimeter, and R is radius in centimeters. In more common units this becomes:

$$P = - \frac{ST}{0.133R} \qquad (2)$$

in which P is now expressed in millimeters of mercury, ST in dynes per centimeter, and R in micrometers.

Another important feature of the fluid in the corners is that, once the equilibrium condition has become established, the radii of the curvature in all the separate corners will be exactly the same if the surface tension is the same everywhere. This is illustrated by the equal lengths of all the arrows representing the radii of curvature in figure 1(A). The reason for this is that, if the fluid pressure is more negative in one corner than in others, fluid molecules will be pulled from the other corners all the way across the surfaces of the alveoli by the surface attraction of molecules (surface tension) until the radius of curvature in the first corner becomes greater while the radii of curvature in the other corners become less. In this way each radius becomes equal to all of the others, and it is also a corollary that the negative pressures underneath the fluid surfaces in all these corners will also become equal.

Weibel and Bachofen [3] have demonstrated that fluid also accumulates in the small angles alongside the capillary bulges in the alveolar walls. This is illustrated in figure 1(B). Where the fluid surface curves over the outside of the capillary bulge the surface tension actually compresses the capillary and causes it to assume an oval shape. On the other hand, in the angles adjacent to the capillaries, the fluid has a concave surface, and the surface tension pulls the fluid away from the alveolar wall. Therefore, the fluid underneath this surface exhibits the same degree of negative pressure as that in the alveolar corners, and the same equations as those given above also apply.

Negative Pressures in Alveolar Apices, Pores, and Flooded Alveoli

Figure 2 illustrates fluid that has accumulated in an apex of an alveolus (where three alveolar surfaces come together), in an alveolar wall pore, and in two separate flooded alveoli. In all of these, the fluid surface is curved in two dimensions, having a spherical surface instead of the cylindrical surface found in corners. Because of this double curvature, the perpendicular force exerted by the surface tension is twice as great as in corners and capillary angles because of the two-dimension pull rather than one dimension.

Thus, note in figure 2(A) the double curvature (spheroid curvature) of the fluid in an apex, and in figure 2(B) the cup-like curvature of the fluid in a pore. In a flooded alveolus, any air that is left within the alveolus assumes the shape of a bubble as shown in figure 2(C), which also has a curvature of a sphere rather than of a cylinder. Therefore, in all of these instances the negative pressure (with respect to alveolar air pressure) obeys the following equation:

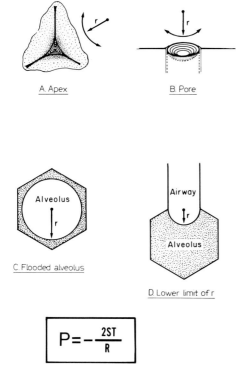

Fig. 2. Accumulation of fluid in alveolar apices (A), alveolar pores (B), and flooded alveoli (C). The air-fluid interface assumes a spherical surface (D).

$$P = - \frac{2ST}{R} \qquad (3)$$

in which P is pressure in dynes per square centimeter, ST is surface tension in dynes per centimeter, and R is radius in centimeters. When this is expressed in more convenient units for use in alveolar and airway calculations, the equation becomes:

$$P = - \frac{ST}{0.0665R} \qquad (4)$$

in which P is pressure in millimeters of mercury, ST is surface tension in dynes per centimeter, and R is radius in micrometers.

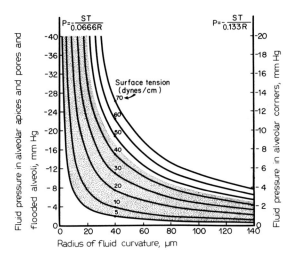

Fig. 3. Effect of different fluid surface tensions and radii of curvature on the negative fluid pressures that occur in alveolar apices, alveolar pores, flooded alveoli, and alveolar corners. Note that the pressures in the corners, because of the cylindrical surface, are only one-half as great as in the apices, pores, and flooded alveoli.

Thus, note that in apices, pores, and flooded alveoli the negative pressure beneath the fluid surface for any given radius of curvature is twice as great as in the alveolar corners and angles. In pulmonary edema, in which alveoli become flooded or partially flooded, the negative pressure, therefore, is calculated according to this spheroid curvature formula, which makes the problem of surface tension in the pulmonary edema process twice as serious as would be true if we were dealing only with fluid in alveolar corners.

Lower Limit of the Radius of Curvature and of the Negative Pressure that Develops in Flooded Alveoli.

If the airway to a flooded alveolus is not blocked, the radius of curvature of the fluid surface can never become less than the radius of the airway. This is illustrated in figure 2(D). If the amount of fluid in the alveolus ever becomes greater than that shown in the figure, then the fluid simply moves up the airway, and the radius of curvature still remains the radius of the airway. It also follows that the amount of positive air pressure that is required to refill a totally flooded lung area is inversely proportional to the radius of the smallest airway subserving the area.

Some Representative Negative Pressures in Alveolar Fluid

Figure 3 illustrates a family of curves showing the negative pressures that will occur beneath the alveolar fluid surfaces for different radii of curvature and different surface tensions. To the right is shown the scale of negative pressures that will occur in corners and in the angles alongside the capillaries while to the left is the scale of pressures (twice as great) that will occur in alveolar apices, alveolar pores, and flooded alveoli. The uppermost curve of this figure represents the negative pressures that occur when the surface tension is approximately equal to that of water at body temperature, 70 dyn/cm. The lowermost curve is approximately that which might occur at the lowest surface tension in the presence of maximal amounts of surfactant. The shaded area illustrates the usual range of surface tensions for normal alveolar fluid with normal amounts of surfactant. Note especially that for normal alveoli with radii of curvature of the corners and apices of only a few micrometers and with normal amounts of surfactant, the corner and apical pressures will range between approximately –5 and –12 mm Hg.

Equilibration of Pressures in All the Fluid Pools of the Alveoli

In the above discussion of fluid in alveolar corners, it was pointed out that the surface tension over the alveolar surface will literally cause movement of fluid from one corner to another in an attempt to equilibrate the pressures everywhere within the alveolus. Thus, actual convective flow of fluid can occur along the surfaces of the alveoli, governed by the pressure differentials within the alveoli.

Figure 4 illustrates several examples of equilibration of pressures within alveoli and even from one alveolus to another. Thus, in figure 4(A), even though the degrees of acuteness of the corner angles within a given alveolus are quite different, the pressures become the same, and the radii of curvature of the fluid surfaces also become the same. To achieve this, the amount of fluid in the acutely angulated corners becomes considerably greater than in the more obtuse corners. But in all instances the pressures equilibrate.

In figure 4(B) it is shown that the pressures in pores and holes will also equilibrate. Note, however, that for a given pressure the radii of curvature in the pores and holes is two times as great as the radii of curvature in the corners. This is represented in this figure by the expression $R = 2r$.

Figure 4(C) illustrates an alveolar wall, showing equilibration of pressures between fluid in a capillary angle, in a hole or a pore of the alveolar

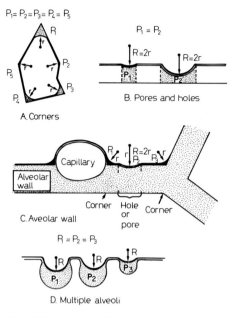

Fig. 4. Transmission of pressures from corner to corner (A), pore to pore (B), corners to pores (C), and alveoli to alveoli (D) as a result of surface transfer of fluid molecules.

wall, and in the corner of the alveolus. Note once again that the radius of curvature in the hole or pore is twice as great as that in the corner or capillary angle because of the spherical nature of the fluid surface in the hole or pore.

Finally, note in figure 4(D) the equilibration of radii of curvature and of pressures from one alveolus to another when they are close enough to each other so that fluid can flow along the surfaces from one alveolus to another. Note also that the smallest alveolus fills more completely with fluid than do others because the radius of its opening is smaller than the others so that the air bubble can protrude less deeply into this alveolus than into the others.

The Alveolar 'Wash' Mechanism

Many different studies have now suggested that interstitial fluid pressures in the junctional tissues of the lungs – the tissues surrounding the

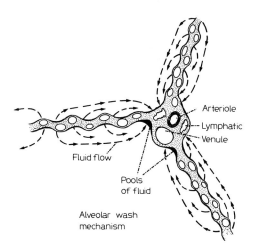

Arteriole

Lymphatic

Venule

Fluid flow

Pools
of fluid

Alveolar wash
mechanism

Fig. 5. The 'alveolar wash mechanism' in which fluid is postulated to flow outward through the epithelium of the alveolar wall, then along the alveolar surface toward the alveolar corner, finally to be absorbed into a junctional space of the lung.

small arterioles, venules, lymphatics, and bronchioles – is almost certainly subatmospheric (probably with pressures equal to about –9 mm Hg). Furthermore, the alveoli are known to drain almost entirely into these junctional tissues. From the previous discussions in this paper, we can postulate that fluid can move in two ways from the alveolar walls to the junctional tissues, (1) movement within the interstitial space of the alveolar wall itself, and (2) movement along the alveolar epithelial surfaces. That is, since the natural drainage is toward the junctional tissues, it is reasonable to assume that the pressures are more negative in the junctional tissues than in the alveolar walls. And, likewise, because of holes in the alveolar corner epithelial membrane, the fluid pressures in the alveolar corners must also become almost equally as negative as the pressures in the junctional spaces. Now, to complete the picture, the surface tension will cause fluid to move along the alveolar surfaces toward the corners, thus creating negative pressures in the pores of the alveolar walls and literally pulling fluid out of the walls into the alveoli. Therefore, net flow would occur outward through the alveolar wall into the surface lining fluid, then along the surface of the alveolus toward the alveolar corners, and finally back through the epithelium into the junctional tissue. Such a mechanism also fits with the observed fact that substances placed in the alveoli, such as albumin solutions, particulate matter, dyes, and so forth, are all transported toward the junctional tissue and thence mainly into the lymphatics [4].

This presumed flow of fluid out of the alveolar wall onto the alveolar surface and thence to the corners and junctional tissues is illustrated diagrammatically in figure 5. It is a mechanism that can be postulated from the basic physics of surface tension phenomena and from known facts about the physiology of fluid in the alveoli and pulmonary interstitium. This mechanism could well be called the *alveolar wash mechanism*. It would be a valuable adjunct to the ciliary mechanism in the respiratory airways for cleansing the lungs.

Mechanism of Alveolar Flooding and the Role of Surface Tension

Balance of Pressures between Alveolar Fluid and Pulmonary Interstitial Fluid as the Determinant of Alveolar Flooding

The quantity of fluid that normally exists in the alveoli is determined by the balance between the rate of fluid entry into the alveoli through the respiratory epithelial membrane versus the rate of fluid leaving through the same membrane. And since the quantity of fluid in the alveoli in any steady-state condition remains constant, the forces moving fluid into the alveoli must equal the forces moving fluid out of the alveoli. It has already been pointed out above that more fluid may be entering than leaving the alveoli through some surfaces such as the alveolar walls, while more fluid might be leaving than entering the alveoli through other surfaces such as in the alveolar corners. But, on the average, these two effects are equal and opposite to each other. Therefore it is possible to analyze alveolar flooding on the basis of relative differences between the *average* alveolar fluid pressure and the *average* pulmonary interstitial fluid pressure.

Several different estimations of the pulmonary interstitial fluid pressure suggest that it ranges between −6 and −12 mm Hg. The values that the authors trust the most are those that we have measured using an alveolar flooding technique [2]. This has been achieved by introducing a catheter into a distal airway, then occluding the airway either by inflating a balloon or by surgically tying the catheter in the airway, next filling the alveoli beyond the catheter with fluid, and finally measuring the equilibrium pressure that develops. Since in the equilibrium state no fluid moves into or out of the alveoli, the pressures on the two sides of the respiratory membrane pores should be equal. The equilibrium pressures that have been measured by this alveolar flooding technique in the normal dog lung at the level of the heart have averaged between –8 and –9 mm Hg. Therefore, the pulmonary interstitial fluid pressure must provide an absorptive force equal to this equilibrium pressure, averaging also about –8 to –9 mm Hg.

Under normal conditions one would not expect to find a significant amount of fluid in the alveoli because of this normal absorptive effect of the negative pulmonary interstitial fluid pressure of –8 to –9 mm Hg. Yet, referring again to figure 3, one sees that the surface tension effect in alveoli lined with water instead of with normal alveolar fluid containing surfactant could easily develop very negative intra-alveolar fluid pressures. The greatest radius of curvature that could obtain for fluid in an alveolus is equal to the radius of the largest sphere that can fit in the alveolus. Therefore, for alveoli with no surfactant and with radii of approximately 130 µm, the surface tension forces would develop intra-alveolar fluid pressures of about –8 to –9 mm Hg, and in all alveoli with radii smaller than 130 µm the pressure would be even more negative than –8 to –9 mm Hg. Consequently, one would expect the alveoli with smaller radii literally to pull fluid out of the interstitium into the alveoli and eventually to become flooded, while those alveoli with larger radii would not fill with fluid.

Thus, figure 6 illustrates a flooded alveolus in the upper part of the figure, a partially flooded alveolus at the bottom, and a dry alveolus to the right. In the flooded alveolus the alveolus is too small to contain an air bubble large enough to prevent the very negative pressures that cause flooding. In the partially flooded alveolus one side of the alveolus has an angle too acute for the surface of the bubble to fit into the corner. Therefore, the surface tension of the bubble has pulled fluid from the interstitium into the corner. But, the alveolus to the right is essentially dry because there are no angles that are acute enough for the surface tension to pull fluid out of the interstitium.

It follows, then, that whether or not an alveolus will become flooded depends upon which is more negative, the fluid pressure in the alveolus caused by the fluid's surface tension or the fluid pressure in the pulmonary interstitium caused by lymphatic drainage and by the balance of forces at the pulmonary microvasculature membranes. If the alveolar pressure is more negative, then flooding will occur.

It is already clear that the degree of negativity of the fluid pressures in the alveoli is a function not only of the radii of curvature of the alveoli but

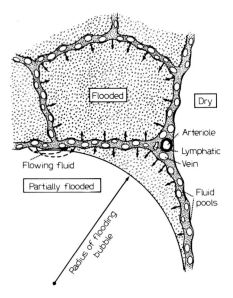

Fig. 6. Flooding of pulmonary alveoli, illustrating that alveoli of very small dimensions become fully flooded while larger alveoli do not. Yet, even large alveoli with sharp corner angulation will become flooded in the corners.

also of the effect of surfactant to decrease surface tension. On the other hand, the degree of negativity of the pulmonary interstitial fluid pressure is determined by the hydrodynamics of the pulmonary interstitium, of the pulmonary lymphatic system, and of the pulmonary vasculature. Such effects as heart failure, lymphatic blockage, reduced plasma colloid osmotic pressure, and increased permeability of the pulmonary capillary membrane can all lead to less negative pulmonary interstitial fluid pressure.

The Interstitial Fluid Pressure at which Alveolar Flooding Will Occur
One can calculate the critical pulmonary interstitial fluid pressure at which one would expect alveolar flooding to occur by first determining the largest air bubble that can be totally contained within the alveolus. (For the average alveolus, approximately the largest bubble that can be completely contained has a radius of about 100 μm.) Then, referring again to figure 3 and using the scale to the left-hand side of the figure and also considering the average surface tension to be 15–20 dyn/cm, one can deter-

mine that the pulmonary interstitial fluid pressure must be at least as negative as –3 mm Hg to prevent alveolar flooding. In a recent study performed by *Parker et al* [2] in our laboratory, the results were consistent with the calculation. The authors increased the interstitial fluid pressure by progressively increasing the vascular pressure, and the pulmonary fluid volume suddenly began to increase when the measured pulmonary interstitial fluid pressure rose only 1 mm or so more positive than the critical value of –3 mm Hg.

However, it should also be kept in mind that, in the absence of surfactant, a much more negative pulmonary interstitial fluid pressure would be required to prevent alveolar flooding. For a maximum alveolar bubble radius of 100 μm and with no surface-active agent in the alveolar fluid, the critical interstitial fluid pressure above which alveolar flooding would occur would be approximately –11 mm Hg. Since the normal pulmonary interstitial fluid pressure probably is not quite this negative – only –8 to –9 mm Hg – one can well understand that alveolar flooding could easily occur in the absence of surfactant even when pulmonary vascular and interstitial fluid dynamics are normal. In fact, recent studies by *Albert et al.* [5] have shown that procedures which remove surfactant from the alveolar fluid do indeed tend to increase the accumulation of fluid within the lungs.

The Transepithelial Route for Alveolar Flooding.

It has often been suggested that alveolar flooding does not occur until the alveolar epithelium ruptures. However, even normally, fluid and its constituents can cross the alveolar membrane, including even large molecules such as albumin [1, 2]. Therefore, there is no reason to invoke rupture of the alveolar epithelium as a necessity for alveolar flooding. On the other hand, if the alveolar epithelium does rupture, alveolar flooding obviously could occur more rapidly. One could easily postulate that the normal presence of negative fluid pressure in the alveolar wall interstitium underneath the alveolar epithelium would normally keep the epithelial cells plastered tightly against their basement membranes. On the other hand, positive fluid pressure underneath the epithelial cells could easily push them away from the basement membrane and cause rupture. Yet, even if rupture did not occur, over a period of time alveolar flooding almost certainly would still occur because of fluid movement through both the small and large holes that normally exist in the alveolar epithelial membrane.

It is also of interest to consider whether severe rupture of the alveolar epithelial membrane without other abnormalities would in itself cause alveolar flooding. The discussions of this paper suggest that physical intergrity of the alveolar epithelium is not necessary to prevent alveolar flooding. Of course, though, rupture of the alveolar epithelium is often associated with other conditions that do cause alveolar flooding. What is important seems to be the balance between alveolar fluid pressure and pulmonary interstitial fluid pressure, the first of these determined by the surface tension in the alveoli and the other by hydrodynamic characteristics of the pulmonary interstitium itself. If the alveolar fluid pressure is more negative than the interstitial fluid pressure, then flooding theoretically will occur with or without alveolar epithelial rupture, though presumably more rapidly if rupture does occur.

References

1   Staub, N.C.; Gee, M., and Vreim, C.: Mechanism of alveolar flooding in acute pulmonary oedema; in CIBA Foundation Symposium 38, Lung Liquids, pp. 255–272 (Elsevier Excerpta Medica, Amsterdam 1976).

2   Parker, J.C.; Guyton, A.C., and Taylor, A.E.: Pulmonary interstitial and capillary pressures estimated from intra-alveolar fluid pressure. J. appl. Physiol. *44:* 267–276 (1978).

3   Weibel, E.R. and Bachofen, H.: Structural design of the alveolar system and fluid exchange; in Fishman and Renkin, Pulmonary edema, pp. 1–20 (American Physiological Society, Bethesda, Ma. 1979).

4   Staub, N.C.: Pulmonary edema. Physiol. Rev. *54:*678–811, (1974).

5   Albert, R.K.; Lakshminarayan, S.; Hildebrandt, J.; Kirk, W., and Butler, J.: Increased surface tension favors pulmonary edema formation in anesthetized dogs' lungs. J. clin. Invest. *63:*1015–1018 (1979).

A.C. Guyton, MD, Department of Physiology and Biophysics, University of Missisippi Medical Center, 2500 N. State St., Jackson, MS 39216 (USA)

Discussion

Clements was surprised about the reflection coefficient of 0.2 for protein across the alveolar epithelium, because commonly the alveolar epithelium is considered tight from electron microscopic studies, but also from other studies, so that many people calculate the reflection coefficient above 0.8. *Guyton* showed that he had calculated this low reflection coefficient, as it can be done from many other studies. *Clements* asked *Reifenrath* about albumin concentration in the air-filled lung estimated by micropuncture and *Reifenrath* said that in his experiments the albumin concentration was independent of the volume of liquid he had obtained. The amount of albumim or other plasma proteins was dependent upon the technique of obtaining the lavage and number of rinsing cycles. After perfusion of the circulatory system, the albumim concentration decreased within the extracellular lining layer. *Von Wichert* asked about the role of the positive-end-expiratory pressure in artificial ventilation. He supposed that by increasing the functional residual capacity the increase in alveolar size brought the surface tension to the right side of the balance between surface forces and interstitial forces. *Guyton* agreed that this would be exactly what he predicted as the really effective value of positive pressure ventilation.

Kapancy pointed out that in the alveolar corners one may have a much larger layer of surface-active material. He asked if *Guyton* thought that this thick layer of surface-active material might explain the phenomenon that at the corners there is a very fast resorption and at the alveolar walls liquid comes out from the tissue. *Guyton* agreed. *Morley* suggested that, as a result of the negative interstitial pressure, the fluid should be sucked out of the alveoli. It may be that the alveolar surface is virtually dry because of this negative pressure and *Guyton* also believed that alveoli surface may be dry. *Clements* disagreed quite strongly on that point with *Guyton*, considering many investigations showing that there is much more than a monomolecular film of liquid over the alveolar epithelial cells. However, this disagreement seemed to defend on the definition of 'dry' because *Guyton* had not meant to imply a monomolecular film.

Prog. Resp. Res., vol. 15, pp. 76–77 (Karger, Basel 1981)

Surfactant Properties in Healthy Men and in Cases of Shock Lung

O.K. Reiss and T.L. Petty

Manuscript not submitted.

Discussion

Rüfer asked if the material purified by *Reiss* had glycohydrates and if staining with PAS was only slight. *Robertson* wondered about the concept of purification and how the substance was defined before purifying. *Reiss* answered that the aim was to purify those substances which were bound tightly without using chemical techniques but by using only strictly physical techniques and that the material one gets in this way has a very constant composition. Concerning a question by *von Wichert, Reiss* said that there are postmortal changes particularly before getting the alveolar macrophages out. In such conditions there is more lysolecithin, normally less than 1% of the lipid phase which is otherwise 2 or 3%. There are no data available about postmortal changes in proteins. *van Golde* asked about the possibility of analyzing lamellar bodies isolated from lungs and *Reiss* said that the problem is to identify lamellar bodies from humans because lamellar bodies from humans, as shown by *Weibel* and colleagues, look entirely different when compared with those from animals. The latter look as though they are partly decomposed so there are many methodological problems. As *Kapancy* pointed out, the reason is that only intravascular perfusion leads to good preservation and if one fixes the lungs from animals and men in the same way, the lamellar bodies look alike. *Reiss* answering a question by *Clements* said that if one plots the isopycnic density of the highly purified material against the alveolar diameter then one gets a progression line, in other words, the larger the alveoli the more protein in the purified surfactant. It appears that the more protein the higher the compressibility of the film. *Reifenrath* raised a more general question, saying that one has to know what surfactant is before starting to purify surfactant and he also mentioned that the material when not highly purified was superb in the kinetic of film-formation to the lowest values of surface tension, so some physiological parameters as the kinetic of film-formation, for instance, are different in purified and nonpurifed fractions. *Reiss* answered that the zonal purified material has a higher saturated lecithin content and somewhat lower cholesterin content than the unpurified fraction, but one of the interesting observations is that when he purified that material he obtained definitely different compositions depending on the size of the alveoli. The general question whether there is any reason to purify surfactant material and if so to what extent it should be done, on the other hand, is not completely answered.

It seemed to *Clements* that one makes a mistake when one starts out by saying *the* surfactant because it is a system, it has many parts and one does not know that the composition will be the same at different stages of its life cycles. What one calls the surfactant when it is being synthesized in the endoplasmatic reticulum may be one thing with whatever proportions of lipids may exist there; a protein component has presumably to be added. The composition is then different and the material must be assembled, packed and stored within the cell, its composition may again be different and its composition is as different, presumably, as its physical property. Its surface activity would be different if we could only attain those forms by fractionation. When the material is extruded from the cells, it goes into an extracellular medium and may bind, for example, calcium, which it probably does not have in the cells (because of the very low calcium concentration in most cells); its density probably changes – it undergoes a structural change, we think to tubular myelin, and then sets up the surface film. Evidence from microscopic studies with flurorocarbonate droplets shows that despite the increases and decreases of lung volume the surface is very stable. The temperature studies we have suggested point out that the interphase is filled with something which is almost pure dipalmitate phosphatidylcholine, and also that other components which we think we see as reproducible in highly purified surfactant fractions are not there, at least as far as one can tell from the zonal dependence of the stress-strain relationship of the lung, and that after that some of the material goes to the subphase. Not imagining that it is vaporized into the air, what we used to call surfactant is now separated and part of it is in the subphase and part of it in the interphase: which is then the true surfactant? After that the surfactant material has to be cleared. The important criteria for its physical properties and composition at that point may be different because it must presumably give a signal to the epithelial cells which tell the cells how to deal with it or how to remove it, depending upon the physiological stage. It is very dangerous to oversimplify the surfactant problems. To try to say we know exactly what compositional and physical characteristics surfactant should have when we see it in the lamellar body or when we wash out the lungs then mix two, three or four physiological compartments, is too much. *Clements* thinks that is the state of our knowledge. *Reiss* agreed with *Clements* statement and added the observation that pure lamellar bodies on the Wilhelmy-balance do not release the material until ions are added. Calcium and to a lesser extent magnesium are required for filmformation. This may be the reason for different results in tension measurements as discussed earlier by *Reifenrath*. Calcium was removed from the highly purified material. *Morley* asked if the crude surfactant might contain lamellar body particles because he felt that just the particles are vital for surfactant spreading and *Reiss* confirmed that he did not have particles in his preparation.

Tierney asked why the minimum surface tension in minced lung material was lower than in lung wash. *Reiss* answered that the reason for this was a different basis for calculation of lipid phosphorus in the shock lung and wet weight in the normal lungs. *Clements* asked if it is possible to exclude postmortal changes in normal lungs to obtain lavage material from lobectomy or pneumonic specimens, but *Reiss* answered that there was not much difference between these specimens, because transplant donor lungs look very fresh indeed. *Morley* thought that these differences came from a different content of lamellar bodies, but *Reiss* answered that it is very difficult, for technical reasons, to make preparations of lamellar bodies in any other species than the rat.

Experimental Findings

Prog. Resp. Res., vol. 15, pp. 78–83 (Karger, Basel 1981)

Control of Surfactant Secretion in Fetal and Newborn Rabbits[1]

Anthony Corbet

Department of Pediatrics, Baylor College of Medicine, Houston, Tex.

Introduction

Goldenberg et al. [10] suggested that pilocarpine given to adult rats caused secretion of lamellar bodies from granular pneumocytes. Since pilocarpine has both cholinergic and adrenergic activities, it was investigated whether surfactant secretion is under control of the sympathetic or parasympathetic nervous system.

Materials and Methods

Fetal and newborn white New Zealand rabbits, gestational age known to within 2 h, were used.

In a first series of experiments, does of 27.5 days gestation were anesthetized with methoxyflurane, laparotomy was performed and the fetal rabbits given intraperitoneal injections of a drug or saline (0.1 ml). In some cases the drug was also administered to the doe through an ear vein (table I). Where appropriate, 2 drugs were given at the same time, an agonist with a specific antagonist. Each fetal rabbit in each litter received the same injection, other litters being used as controls. Because rabbits produce atropinesterase, atropine was given in large doses sufficient to maintain paralysis of the maternal pupillary light reflex [2].

After drug administration, the abdominal incision was closed and the doe allowed to recover for 2.5 h, when the fetal rabbits were delivered and killed immediately by injection with pentobarbital.

After tracheostomy and insertion of a tracheal tube, a pressure-volume curve or lung lavage by standard methods [7] was performed. The deflation limb of the pressure-volume diagram, normalized as a percentage of total lung capacity (TLC), was used for analysis, and

[1] Supported by US Public Health Service grant RR-05425.

Table I. Drugs injected in maternal and fetal rabbits

	Fetal dose	Maternal dose
Pilocarpine, mg	5	–
Atropine, mg	10	30–90
Propranolol, mg	0.1	4.5
Phenoxybenzamine, mg	–	10
Muscarine, μg	0.3	100
Isoxsuprine, mg	0.5	–
Phenylephrine, mg	1	–
Aminophylline, mg	5	–

Table II. Data for lung water, pressure-volume curves and total phospholipid recovered by lung lavage in fetal rabbits of 27.5 days gestation injected with specific agonist and antagonist drugs (values are mean ± SE)

Injection	Litters	% H_2O lung	% TLC-5	TPL μg/g DLW
Saline	6	89 ± 0	25 ± 2	32 ± 3
Pilocarpine	6	90 ± 1	43 ± 4[1]	74 ± 6[1]
Pilocarpine, atropine	4	89 ± 0	25 ± 3	
Pilocarpine, propranolol	4	89 ± 1	32 ± 2	31 ± 3
Pilocarpine, phenoxybenzamine	6	91 ± 1	24 ± 2	
Muscarine	5	87 ± 1	31 ± 2	
Isoxsuprine	10	85 ± 1[1]	42 ± 2[1]	52 ± 4[1]
Isoxsuprine, propranolol	4	87 ± 0	19 ± 2	
Isoxsuprine, phenoxybenzamine	4	85 ± 1[1]	43 ± 3[1]	
Phenylephrine	3	85 ± 0[1]	30 ± 3	
Aminophylline	5	89 ± 1	36 ± 2[1]	62 ± 6[1]

[1] Significantly different from saline controls, $p < 0.05$.

the volume at 5 cm H_2O (% TCL-5) taken as an index of surfactant activity *in situ*. The lung lavage fluid was analyzed for total phospholipid (TPL) which was taken as a quantitative measure of pulmonary surfactant [4].

In a second series of experiments, does of 29.5 days gestation were used. At laparotomy, intraperitoneal injections of saline, atropine or propranolol were made, and 15 min later the fetal rabbits were delivered. Some litters were killed immediately at birth, and some were allowed to breath for 45 min before sacrifice. Tracheostomy and lung lavage were performed and measurements made of TPL in lavage fluid [6].

After completion of these procedures the dissected lungs were weighed, dessicated, dry lung weight (DLW) measured, and the percentage lung water (% H_2O) calculated.

Table III. Total phospholipid (µg/g DLW) recovered in lung lavage of newborn rabbits, 29.5 days gestation, injected 15 min before birth, and either killed at birth or allowed to breath for 45 min (values are mean ± SE)

Injection	Nonbreathing	Breathing
Saline	191 ± 22	293 ± 34[1]
Propranolol	167 ± 20	244 ± 28[1]
Atropine	226 ± 29	217 ± 29

[1] Significantly different from nonbreathing controls, p < 0.05.

Results

Litters of 27.5 days gestation injected with pilocarpine, isoxsuprine or aminophylline showed a significant increase in the volume of air retained on deflation from TLC to 5 cm H_2O when compared with litters injected with saline (table II). The effects of pilocarpine were blocked with atropine, phenoxybenzamine and propranolol, and the effects of isoxsuprine were blocked with propranolol but not with phenoxybenzamine. Atropine, propranolol and phenoxybenzamine, given alone, and muscarine and phenylephrine had no effect on the pressure-volume diagram. Both isoxsuprine and phenylephrine produced a reduction in lung water.

Litters of 27.5 days gestation injected with pilocarpine, isoxsuprine or aminophyllin showed a significant increase in TPL recovered by lung lavage. This effect of pilocarpine was blocked with propranolol (table II).

Does injected with pilocarpine or muscarine salivated profusely. This effect of pilocarpine was blocked by atropine and phenoxybenzamine, but not by propranolol.

Litters of 29.5 days gestation injected with saline or propranolol, and breathing for 45 min, showed a significant increase of TPL recovered by lung lavage when compared with similarly injected nonbreathing litters, but atropine prevented this increase (table III).

Discussion

Increased retention of air on deflation to low transpulmonary pressures is thought to indicate decreased surface tension from increased surfactant secretion induced with pilocarpine, isoxsuprine and aminophylline [4, 7, 8]. If airway closure was responsible for increased retention it should produce a reduction in volume extracted from the lung between 5 and 0 cm H_2O, but this was never the case [4, 7, 8]. Fetal lung tissue shows plastic rather than elastic behavior, so tissue recoil forces at low volumes are absent and do not explain increased deflation stability [7]. Dehydration of the lung may concentrate surfactant, but if this explains increased stability in isoxsuprine-treated lungs, similar effects would be expected in those treated with phenylephrine. Configurational changes are possible but are considered very unlikely [7].

Increased TPL recovered by lung lavage from litters treated with pilo-carpine, isoxsuprine or aminophylline (table II) confirms the above inter-pretation of the pressure-volume curves.

Inhibition with atropine, isoxsuprine and phenoxybenzamine ot the effect of pilocarpine on the pressure-volume curve (table II), and inhibiton with propranolol of the effect of pilocarpine on TPL recovered by lung lavage (table II), suggests that pilocarpine exerts its effect by releasing adrenergic mediators through stimulation of the adrenal medulla, sympa-thetic ganglia or possibly storage sites in the lung. Atropine is known to block the effect of pilocarpine on the adrenal medulla and sympathetic ganglia [12]. The effect with isoxsuprine and the absence of effect with muscarine (table II) suggests that adrenergic, but not cholinergic, receptors are present at the granular pneumocyte. The blocking effect of propranolol and the lack of effect of phenylephrine suggests that these receptors are β-adrenergic, not α-adrenergic. The blocking activity of phenoxybenzamine may have been due to its atropine-like properties, since stimulation of sali-vation by pilocarpine, known to be cholinergic, was blocked with phen-oxybenzamine. Thus, these experiments suggest that TPL secretion in the lung is mediated by β-adrenergic receptors. Since these receptors are known to control the intracellular levels of cyclic-AMP [3], and amino-phylline, a phosphodiesterase inhibitor, is known to increase cyclic-AMP, the results of our experiments with aminophylline (table II) are consistent with this hypothesis [4].

Dobbs and Mason [9] have demonstrated β-adrenergic, but not chol-inergic, effects on granular pneumocytes grown in tissue culture. *Abdellatif and Hollingsworth* [1] could not observe muscarinic stimulation of TPL secretion in isolated perfused lung, but could observe such activity in intact fetal rabbits.

Because of the evidence for β-adrenergic receptors it was surprising that TPL secretion at birth was blocked by atropine and not by propanolol (table III) [5]. The doses of atropine used were large, but because rabbits produce atropinesterase, large doses of atropine were required to maintain inhibition of the maternal pupillary light reflex [5]. However, *Lawson and Huang* [11] have also shown that surfactant secretion at birth is blocked by atropine, but not by propranolol.

It is concluded that control of surfactant secretion is complex. In addition to β-adrenergic mediators, there must be other mediators, e.g., prostaglandins, which may be under cholinergic control.

Summary

Fetal rabbits of 27.5 days gestation were injected with pilocarpine, isoxsuprine, aminophylline, muscarine and phenylephrine and with specific antagonist drugs atropine, propranolol and phenoxybenzamine. Pilocarpine, isoxsuprine and aminophylline produced increased deflation stability in static pressure-volume curves performed on the lungs, and this was reflected in increased total phospholipid recovered by lung lavage. The effects of pilocarpine were blocked by atropine, phenoxybenzamine and propranolol, and it was concluded that pilocarpine exerted its effect through stimulation of the sympathetic nervous system. This was supported by similar effects produced with isoxsuprine and aminophylline, but not with muscarine. However, atropine but not propranolol blocked the secretion of lung phospholipid with the onset of breathing at birth in fetal rabbits of 29.5 days gestation. This suggests the importance also of the parasympathetic nervous system.

References

1   Abdellatif, M. M. and Hollingsworth, M.: The *in vitro* and *in vivo* effects of oxotremorine on the phosphatidylcholine content of washes of neonatal rabbit lungs. Br. J. Pharmacol. *61*: 502P (1977).

2   Ambache, N.; Kavanagh, L., and Shapiro, D. W.: A rapid procedure for the selection of atropinesterase-free rabbits. J. Physiol., Lond. *171*: 1P (1964).

3   Butcher, R. W. and Sutherland, E. W.: Adenosine 3',5'-monophosphate in biological materials. J. biol. Chem. *237*: 1244–1288 (1962).

4   Corbet, A.J.S.; Flax, P.; Alston, C., and Rudolph, A. J.: Effect of aminophyllin and dexamethasone on secretion of pulmonary surfactant in fetal rabbits. Pediat. Res. *12*: 797–800 (1978).

5   Corbet, A.J.S.; Flax, P.; Alston, C., and Rudolph, A.J.: Effect of isoxsuprine on secretion of pulmonary phospholipids in fetal rabbits. Aust. paediat. J. *15*: 7–9 (1979).

6   Corbet, A.J.S.; Flax, P.; Alston, C., and Rudolph, A. J.: Role of autonomic nervous system controlling secretion of pulmonary phospholipids in the rabbit at birth. Aust. paediat. J. *15*: 238–242 (1979).

7   Corbet, A.J.S.; Flax, P., and Rudolph, A. J.: Reduced surface tension in lungs of fetal
    rabbits injected with pilocarpine. J. appl. Physiol. *41*: 7–14 (1976).
8   Corbet, A.J.S.; Flax, P., and Rudolph, A. J.: Role of autonomic nervous system
    controlling surface tension in fetal rabbit lungs. J. appl. Physiol. *43*: 1039–1045 (1977).
9   Dobbs, L. G. and Mason, R. J.: Pulmonary alveolar type II cells isolated from rats.
    Release of phosphatidylcholine in response to beta-adrenergic stimulation. J. clin.
    Invest. *63*: 378-387 (1979).
10 Goldenberg, V. E.; Buckingham, S., and Sommers, S. C.: Pilocarpine stimulation of
    granular pneumocyte secretion. Lab. Invest. *20*: 147-158 (1969).
11 Lawson, E. E. and Huang, H. S.: Neurogenic mediation of augmented surfactant secre-
    tion at birth. Abstract. Pediat. Res. *12*: 564 (1978).
12 Root, M. A.: Certain aspects of the vasopressor action of pilocarpine. J. Pharmac. exp.
    Ther. *101*: 125–131 (1951).

A. Corbet, MD, Department of Pediatrics, Baylor College of Medicine, Houston, TX
77030 (USA)

Discussion

Corbet in reply to a question of *Robertson*, said that anything that stimulates the sympathetic nervous system, whether it be alpha or beta reduces the lung water content in his experiments, for this phenomenon various explanations have been given, but the changes of the
phospholipid suggests strongly that a change of the surface tension properties takes place.

In answering a question by *Morley*, *Corbet* said that atropine also has a demonstrable
effect, but there is no reason to translate these data directly to clinical problems. *Robertson*,
asking about the role of bronchoconstriction induced pharmacologically was answered that
a bronchoconstriction produced a change in saline pressure volume curve but not in the air
pressure volume curve. *Clements* suggests that there may be differences in the action of beta
agents, especially beta II agents are very effective, and *Corbet* agreed. *Clements* thinks that the
relative secretory and cardiovasculary effects of the sympatheticomimetic agents are different
and it might be necessary to stimulate the epithelium of the lung to facilitate secretion perhaps
more than the animals could tolerate.

Prog. Resp. Res., vol. 15, pp. 84–92 (Karger, Basel 1981)

Biochemistry of Lung Surfactant: Apoprotein-Phospholipid Interaction[1]

K.S. Zänker, V. Breuninger, D. Hegner, G. Blümel, and J. Probst[2]

Institut für Experimentelle Chirurgie der Technischen Universität München; Institut für Pharmakologie, FB Tiermedizin, Ludwig-Maximilians Universität, München, and BG-Unfallklinik, Murnau

Introduction

The pulmonary surfactant lining the alveoli of mammalian lungs is supposed to be a lipid-protein complex, the major constituent of which is dipalmitoylphosphatidylcholine [1, 3]. Advances in the study of pulmonary surfactant accumulated data on apoprotein (s) found with the surfactant lipids in alveolar lavage. The metabolism of the apoproteins has been studied by *King et al.* [4] and binding properties of lung lavage protein(s) to phosphatidylcholine have been evaluated recently [8]. However, whether lung surfactant contains specific apoprotein(s), which are not serum proteins, has not been established unequivocally [5].

The working hypothesis that specific lung proteins are associated with surfactant lipids was investigated with biochemical, immunological and physicochemical methods and the results, obtained from highly purified surfactant apoprotein(s), are the subject of this report.

Material and Methods

Purification of Surfactant Apoproteins

Surfactant apoprotein(s) were isolated and purified from rat lung lavages by the method described previously [9]. In brief, the purification procedure was carried out by ammonium sulfate precipitation steps (40% and 60% saturated [$(NH_4)_2SO_4$]) and sucrose density centrifugation. This purification scheme turned out to be a powerful tool to gain a rapid and acceptable yield of surfactant apoprotein(s) preparation (purification factor $\sim$ ×37).

[1] Supported in part by the 'Deutsche Forschungsgemeinschaft', Bonn-Bad Godesberg.
[2] *K.S.Z.* is a recipient of a grant from the 'Hauptverband der gewerblichen Berufsgenossenschaften'. Skillful technical assistance of Mrs. *E. Sincini* is acknowledged.

Preparation of Surfactant Antiserum

2 rabbits were inoculated with 2 ml of antigen preparation (protein of the final purifi-cation step, 0.59 mg/ml), mixed with 1.5 ml of complete Freund's adjuvant. The dose was quartered and given intradermally to each limb. Subsequent inoculations of antigen prepara-tions were made once a week for a total of 6 weeks; the boosters were done without Freund's adjuvant. The globulin fraction of the rabbit serum was separated according to *Goldman* [2] and taken for immunodiffusion experiments, described by *Ouchterlony* [7].

Sucrose Density Centrifugation of the Apoprotein-Phospholipid Complex

Linear 5–20% (w/v) sucrose gradients in 50 mM Tris-HCl buffer, pH 7.5 at 20°C were prepared in polycarbonate tubes. On top of the 12.2 ml gradients, apoprotein solutions were incubated for 30 min with ^{14}C-methyl-phosphatidylcholine (^{14}C-PC; spec. activity 0.01 mCi/mmol phosphorus; New England Nuclear, Boston, Mass.) and afterwards centri-fuged at 38,000 rpm for 16 h in a Beckmann SW-40 rotor. About 28 fractions were collected from the bottom of the gradient and 200-µl aliquots were tested in a scintillation counter (Berthold & Frieseke, FRG) for ^{14}C-PC, bound to surfactant apoprotein(s).

Spectroscopic Studies

Spectrofluorometry was done with a commercial available device (Perkin-Elmer, MP4), equipped with a xenon lamp. Apparent excitation and emission spectra were scanned from surfactant apoprotein(s), bovine serum albumin and rabbit γ-globulin solutions (50 mM Tris-HCl, pH 7.3 at 20°C, 1 mg protein/ml).

Electron paramagnetic resonance spectra were recorded with a Brucker spectrometer (FRG) equipped with a variable temperature accessory. L-α-lecithin (dipalmitoyl phosphat-idyl choline) having a paramagnetic N-oxyl-4', 4'-di-methyloxazolidine ring at position 10,3 on a palmitic acid chain has been prepared. Suberic acid monoethylester (50 g) and thionyl-chloride (29.8 ml) were heated under reflux for 3 h and then allowed to stand for 2 h at 4°C, the acidic chloride was purified by distillation (162°C, 155 Torr), yielding 14.25 g of acidic chloride. A solution of 42.7 ml octylbromide in 300 ml of dry ether was added in the usual way, to 5.85 g of magnesium turnings in 100 ml of dry diethyl ether. The Grignard solution was cooled in an ice bath and 23.19 g CdCl$_2$ was added in one portion. The ice bath was removed, and after a period of 30 min the ether was almost removed by distillation. Dry benzene (300 ml) was added and 14.25 g of ethyl-7-(chloroformyl-1)-heptanoate-ether in 150 ml of dry benzene was added over a period of 10 min with rapid stirring. The mixture was then heated to reflux for 1 h, after which the mixture was cooled in an ice bath and 50 ml of water was added slowly with stirring. Then a large excess of 0.1 N sulfuric acid was added until two distinct phases were formed. The upper phase was collected and dried over anhydrous sodium sulfate. The benzene was removed under reduced pressure and the solid recrystallized from pentane, yielding 6.4 g of ketopalmitate. To 500 ml of toluene were added 0.1 mol of the ketopalmitate, 100 ml of 1-amino-2-methyl-1-propanol and 100 mg of *p*-toluene-sulfonic acid monohydrate. The mixture was refluxed for 12 days using a Dear-Stark trap for water removal. The toluene phase was then washed with eight 200-ml portions of saturated sodium bicarbonate solution and six 200-ml portions of water and dried with anhydrous sodium sulfate. The 4', 4'-dimethyloxazolidine (1.6 g) was dissolved in 200 ml of diethyl ether, cooled to 0°C in an ice bath and 100 ml of diethyl ether containing 0.972 g of *m*-chloroperbenzoic acid was added over a period of 2 h. The mixture was allowed to stand for 12 h, at which time the ether phase was washed four times with 100 ml of saturated sodium bicarbonate and four

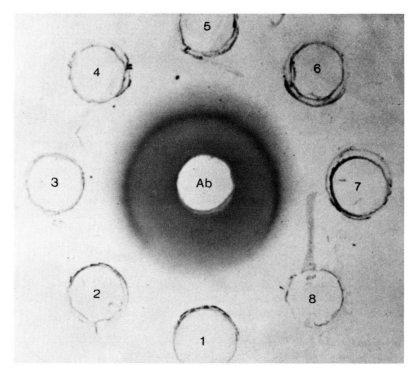

Fig. 1. Immunodiffusion analysis. Central well contained the surfactant antiserum. Other wells had (1) rat serum proteins, and (2–8) different preparation steps of surfactant apoprotein(s) according to *Zänker et al.* [9]. Surfactant apoprotein(s) antiserum react with apoprotein(s) but failed to react with normal rat serum.

times with 100 ml of water. The ether was removed under reduced pressure and the yellow, viscous oil chromatographed on a silica gel column, eluting with benzene-ether (7:3 v/v). The center portion of the fast-moving yellow band was collected, twice recrystallized from pentane, yielding 0.3 g N-oxyl-4', 4'dimethyloxazolidine derivative of the ketopalmitate.

Preparation of Anhydride

To the label, dicyclohexylcarbodiimide, dissolved in 2 ml dry carbon tetrachloride, was added (1:0.8 mmol). After 1 day, the precipitated dicyclohexylurea was removed by filtration and the carbon tetrachloride removed under reduced pressure. The acylation of *L-α*-glycerophosphorylcholine by the label (10,3)-anhydride followed the procedure described by *Robles and Van den Berg* [6]. Final purification of the 10,3 label was done on thin-layer chromatography where the label component behaved like natural lecithin.

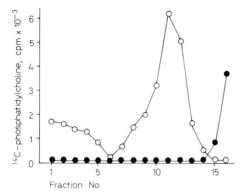

Fig. 2. Sucrose density centrifugation of a phospholipid-apoprotein complex. ¹⁴C-phospholipids can be spun down in the presence of surfactant apoprotein (0); in the absence, however, ¹⁴C-phospholipids remain on top of the gradient (●).

Phospholipid-Apoprotein Interaction, Monitored by ESR Spectra

Phospholipid vesicles, consisting of dipalmitoyl-lecithin and the synthesized label (10,3) or a stearic acid label (12,3) have been formed as multilamellar vesicles and by sonication unilamellar vesicles were obtained. This vesicles were divided into two probes, one of which was incubated for 90 min at 37°C with a highly purified surfactant apoprotein preparation. After the incubation time, apoprotein(s) attached to the vesicles were separated by ultracentrifugation (18,000 rpm at 4°C, 45 min) and subjected to temperature-dependent (~ 10°C/h, 13–55°C) ESR measurements. Control experiments were performed only with vesicles, lacking surfactant apoprotein.

Results

The agar gel immunodiffusion indicated that the IgG fraction of the immuneserum contained antibodies, directed against proteins, derived from different purification steps of rat lung lavage (fig. 1, wells 2–8). No precipitation line was seen against rat serum proteins (well 1). Sucrose density centrifugation of the ¹⁴C-PC-apoprotein complex revealed two peaks of radioactivity (fig. 2). One small peak of activity was located near to the bottom of the centrifugation tube (fraction 1–5), whereas a second, broader peak showed its maximum of radioactivity between fractions 10 and 11. In contrast, pure ¹⁴C-PC remained on top of the gradient (fraction 15, 16).

Table I. Fluorescence characteristics of surfactant apoprotein compared to serum-derived proteins

	Surfactant apoprotein	γ-Globulin	BSA
Emission, nm			
λ_{em}^{max}	335 ± 1	329 ± 1	338
HWB^{em}	64 ± 1	56	58
Excitation, nm			
λ_{exc}^{max}	290	287	282
HWB^{exc}	42	32	30

HWB = Half with breadth

Fluorescence parameters of rabbit γ-globulin, bovine serum albumin and a highly purified surfactant apoprotein preparation were determined in order to give evidence that surfactant apoprotein is not identical with γ-globulin or albumin. Table I summarizes the findings. There is a marked difference between these proteins concerning their maximum wavelengths of excitation and emission. In figures 3 and 4, the functional implications of the synthesized spin label (10,3) are demonstrated, which is in use to investigate the structural changes which are assumed to take place at the phase transition when incubated with surfactant apoprotein(s).

A first series of experiments, using a stearic acid label (12,3), and surfactant apoprotein(s), came out with a loss of the prephase transition of the label.

Discussion

The current debate on the molecular composition of lung surfactant prompted us to have a proper look at the interactions of highly purified protein(s) from rat lung lavage to phosphatidylcholine. Immunodiffusion analysis of the raised antiserum, directed against surfactant apoprotein(s) produced one prominent precipitation line; therefore, it is apparent that the described experiments for the evaluation of apoprotein-phospholipid interaction are carried out with specific alveolar fluid protein(s).

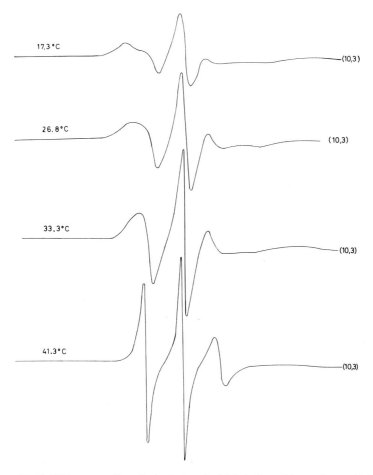

Fig. 3. ESR spectra of synthesized phosphatidyl choline with one fatty acid (palmitic acid) labelled down the chain at 10,3 and incorporated in dipalmitoyl lecithin vesicles. The temperature dependent conformational probabilities can be estimated from the variation in order parameter.

Furthermore, ample evidence that apoprotein-phospholipid complexes are formed *in vitro,* can be given by sucrose density centrifugation. Phospholipids, spun down on a sucrose gradient, banded on the very top of the meniscus. In the presence of surfactant apoprotein(s), however, the phospholipids can be sedimented, suggesting complex formation with surfactant apoprotein(s). Carefully evaluating the preliminary ESR data,

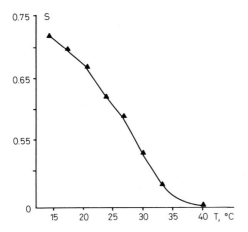

Fig. 4. Plot of the order parameter for phospholipid spin label 10,3 present to the extent of 1 mg in an aqueous dispersion of dipalmitoyllecithin, as a function of temperature.

phospholipid-apoprotein(s) formation is to be confirmed. The disappearance of the prephase transition of the stearic acid label (12,3) also propose a form of phospholipid-apoprotein(s) interaction, the nature of which has to be figured out in the future. An intrinsic feature of spin label experiments is that the molecules used are never identical with the natural compound. To overcome this disadvantage, a variety of strategems were practised in order to substantiate the view that phospholipids interact with proteins, purified from rat lung lavage. The major question, whether these proteins are related to pulmonary surfactant *in vivo,* cannot be answered ultimately. We are sensitive to the idea that the surface activity of lung surfactant is dominated by the lipid and, cautiously deduced from ESR data, apoprotein(s) has little influence on to the gel-to-liquid crystal transitions of the phospholipid. Nevertheless, our results emphasize that there are interactions between phospholipids and surfactant apoprotein(s) and by means of fluorescence and immunological studies, these proteins are *not* identical with serum proteins above all albumin and γ-globulins. The physiological implications of surfactant apoprotein(s), however, is far from clear.

Summary

The interaction of phospholipids with a highly purified surfactant apoprotein preparation from rat lung lavage was investigated by biochemical and biophysical methods. The results indicate that surfactant apoprotein has an affinity *in vitro* to phospholipids and these protein(s) have no immunological counterpart in the serum. Preliminary ESR data on phospholipid-surfactant apoprotein recombination experiments also suggest molecular interaction, but not mainly due to conformational changes of the phospholipid backbone. The protein moiety, however, seems to play an integral part in lung surfactant for normal function.

References

1   Frosolono, M.F.; Charms, B.L., and Slivka, S.: Isolation and characterization and surface chemistry of a surface-active fraction from dog lung. J. Lipid Res. *11:* 439–457 (1970).

2   Goldman, M.: Fluorescent antibody methods; 3rd ed., pp. 87–96 (Academic Press, New York 1968).

3   King, R.J. and Clements, J.A.: Surface active materials from dog lung. II. Composition and physiological correlations. Am. J. Physiol. *223:* 715–726 (1972).

4   King, R.J.; Martin, H.; Mitts, D., and Holmstrom, F.M.: Metabolism of the apoproteins in pulmonary surfactant. J. appl. Physiol. *42:* 483–491 (1977).

5   Maguire, J.J.; Shelley, S.A.; Paciga, J.E., and Balis, J.U.: Isolation and characterization of proteins associated with the lung surfactant system. Prep. Biochem. *7:* 415–425 (1977).

6   Robles, E.C. and Van den Berg, D.: Synthesis of lecithins by acylation of O-(*sn*-glycero-3-phosphoryl)choline with fatty acid anhydrides. Biochim. biophys. Acta *187:* 520–526 (1969).

7   Ouchterlony, Ö.: Diffusion-in-gel methods for immunological analysis II. Progr. Allergy, vol. 6; pp. 30–154 Karger, Basel 1962).

8   Zänker, K.S.; Tölle, W.; Wendt, P., and Probst, J.: On the trace of pulmonary surfactant. Biochem. Med. *20:* 40–53 (1978).

9   Zänker, K.S.; Wendt, P.; Blümel, G., and Probst, J.: Partial purification and characterization of phosphatidyl choline binding proteins from rat lung lavage. Biochem. Med. (in press).

K.S. Zänker, MD, DVM, Institute for Experimental Surgery, Technical University of Munich Medical School, Ismaningerstrasse 22, D-8000 Munich 80 (FRG).

Discussion

Answering a question by *Clements, Zänker* said, he had tried to make the Scatchard plot, but the results of these experiments are not satisfactory. A question by *von Wichert* about how trypsin destroys the protein component was answered in that it is quite evident that trypsin clears the protein, so that some binding sites will disappear. Asked by *Büsing* about the importance of the interaction between plasma proteins and the specific surfactant proteins, *Zänker answered that there is no evidence for an interaction because the purification procedure will rule out plasma proteins. Reifenrath* added a few comments referring to the difference between a biochemical and a physiological point of view. Surface properties are varied to a very large extent by addition of only small amounts of other substances, for instance cholesterol or apoprotein, which change the physical state of phosphatidylcholine. These nonplasma proteins might be important with regard to the kinetics of film formation. One has to distinguish between minimum surface tension and kinetic of film formation. Plasma proteins may alter minimum surface tension but not the kinetics of film formation, which could be shown in amniotic fluid studies.

Prog. Resp. Res., vol. 15, pp. 93–103 (Karger, Basel 1981)

Changes in Pulmonary Surfactant Metabolism and Lung Free Cell Populations from Rats Exposed to Asbestos and Glass Dusts [1]

T.D. Tetley and R.J. Richards

Department of Biochemistry, University College, Cardiff

Introduction

Inhalation of airborne particles often results in an inflammatory response in the lung, depending on the physicochemical nature of the material inhaled and its subsequent interaction with lung components. One of the first biological barriers to harmful deposits in the lung is the alveolar surface lining material, pulmonary surfactant. In this respect *Desai et al.* [1] and *Desai and Richards* [2], have shown that surfactant plays a protective role by rapidly absorbing onto particulate matter, probably in advance of phagocytosis by lung cells, reducing their initial susceptibility to the dusts. Alveolar macrophages also play an important role in lung clearance by phagocytosing deposited materials, before migrating from the lung surface.

The research group at Cardiff is presently concerned with the initial cellular and biochemical changes in the lung following deposition of materials with widely differing properties (e.g. chemistry, shape, density etc.) which can be related to the relative known toxicity of the materials. Clearly, effective removal of toxic agents from the lung is of prime importance, and changes in lung surface lipoproteins and lung free cells will affect this and other lung processes. Consequently, measurement of these two parameters forms a basic part of all our investigations, and parallel studies on surfactant metabolism have helped in our interpretation of these results. The present paper summarises the observations from a number of detailed studies with respect to these lung surface mechanisms following dust inhalation.

[1] The financial support of the Medical Research Council for some of this work is gratefully acknowledged.

Table I. Mean and lavaged lung and body weights of rats exposed to chrysotile and amosite asbestos, fibreglass, powdered glass and the equivalent controls (see 'Materials and Methods' for details)

Interval after initial exposure	Body weight, g		Lung weight, g	
	control	test	control	test
Experiment 1				
3 days	290	318	1.60	1.90
3 weeks	361	350	1.88	2.04
9 weeks	506	453	2.12	2.88
15 weeks	542	493	1.88	2.93
30 weeks	568	457	2.94	3.11
50 weeks	479	565	2.37	4.49
Experiment 2				
Low dose				
18 weeks	492	450	3.02	2.80
30 weeks	526	500	2.73	2.71
High dose				
25 weeks	555	516	2.47	3.18

Exposure conditions	Body weight, g	Lung weight, g
Experiment 3		
Control	635	2.19
Amosite asbestos	619	2.73
Fibreglass	585	2.49
Powdered glass	640	2.05

Materials and Methods

Dust Exposure Conditions

Experiment 1 and 2. Specific pathogen-free (SPF) Wistar rats, 2–6 months of age, were exposed to UICC Rhodesian Chrysotile A in 1.4 m³ inhalation chambers [17] at the Medical Research Council's Pneumoconiosis Unit, Llandough, S. Wales, U.K. Equivalent control rats were kept in the same conditions without dust exposure. The test rats were exposed for 7 h a day, 5 days a week at the following dose levels:

Experiment 1: 15 weeks exposure, total respirable dose 6,503 mg/m³.

Experiment 2: Low dose, 18 weeks' exposure, total respirable dose 883 mg/m³; high dose, 20 weeks' exposure, total respirable dose 7,181 mg/m³.

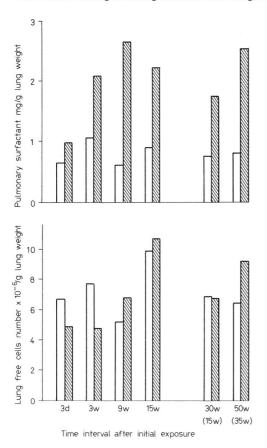

Fig. 1. Experiment 1. The effect of inhalation of high levels (total respirable dose 6,503 mg/m³) of chrysotile asbestos on rat pulmonary surfactant levels and lung free cell numbers. Clearance periods are shown in parentheses. □ = Controls; ▧ = test.

Experiment 3. SPF Sprague-Dawley rats (6–8 weeks of age) were exposed to amosite asbestos, fibreglass or powdered glass in an identical manner to that described above, for a period of 8 weeks. Equivalent control animals were kept for the same period. The total respirable dose levels were as follows: amosite asbestos 7,068 mg/m³; powdered glass 6,654 mg/m³, and fibreglass 6,934 mg/m³. All groups of animals were fed and watered *ad libitum.* At specific time intervals (see below) the animals (6 in each group) were killed using CO_2 asphyxiation and their lungs removed for analysis as previously described [15].

Time Intervals for Analysis
Experiment 1. Analysis during exposure took place 3 days, 3, 9 and 15 weeks after initial exposure. Two groups were exposed for 15 weeks, then allowed 15 and 35 weeks clearance (30 and 50 weeks after initial exposure) before analysis.

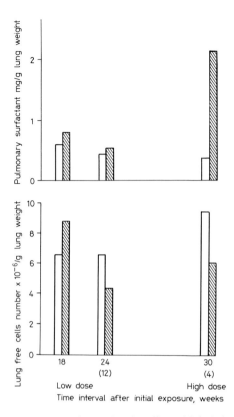

Fig. 2. Experiment 2. The effect of inhalation of low levels (total respirable dose 883 mg/m³) of chysotile asbestos on rat pulmonary surfactant levels and lung free cell numbers. Clearance periods are shown in parentheses. □ = Controls; ▨ = test.

Experiment 2. Low dose: One group of rats was sacrificed immediately following 18 weeks' exposure while another group experienced the same exposure but was allowed 12 weeks' clearance (30 weeks after initial exposure) before analysis. High dose: These animals were exposed for 20 weeks and analysed after 4 weeks' clearance (24 weeks after initial exposure).

Isolation of Pulmonary Surfactant and the Lung Free Cell Population.
Immediately following removal, the lungs were lavaged 10 times with 0.15 M NaCl. The lung free cell number was determined (on pooled lavages) using a Neubaur haemacytometer. The washings from the 6 identically treated animals were then combined and centrifuged at 300 g for 20 min to sediment the lung free cells. The supernatant was centrifuged at 1000 g for 1 h. The resulting pellet was then resuspended in 21% NaCl and centrifuged at 1,500 g for 25 min. The upper pellicle of the final three-phase system was removed, dialysed against H_2O, pH 7.0, for 48 h and lyophilised to give a dry weight [15].

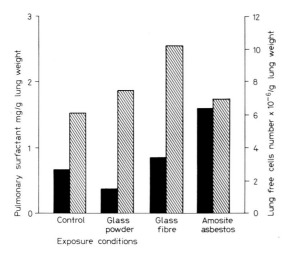

Fig. 3. Experiment 3. The effect of inhalation of amosite asbestos, powdered glass or fibreglass (total respirable doses of 7,068, 6,934 or 6,654 mg/m³, respectively) for 8 weeks on rat pulmonary surfactant (■) and lung free cell numbers (▨).

Staining of Lung Free Cell Populations.

Air-dried smears of the lung free cells were fixed in glutaraldehyde and stained with Giemsa in order to size and determine the incidence of multinucleate cells in experiments 1 and 2 only. Slides were prepared in an identical manner from lavages obtained in experiment 3.

Results

While exposure to asbestos did not cause a significant change in body weight, there was a gradual increase in lung weight following inhalation of high doses of this material, shown in table I, which agrees with previous studies following silica and asbestos exposure [7, 15, 16]. Inhalation of fibreglass also appeared to cause a small increase in lung weight (table I).

Inhalation of high doses of chrysotile induced increased pulmonary surfactant levels within 3 weeks of exposure (2 times control level, fig. 1), which remained elevated during exposure (2–4 times control level, fig. 1) and even after removal from the dust stimulus (2–5 times control level, fig. 1, 2). In contrast, low doses of chrysotile did not produce this marked response (fig. 2). Exposure to high doses of amosite asbestos also resulted in elevated surfactant levels (over twice those of control, fig. 3), although the equivalent dose of fibreglass or powdered glass caused either no change

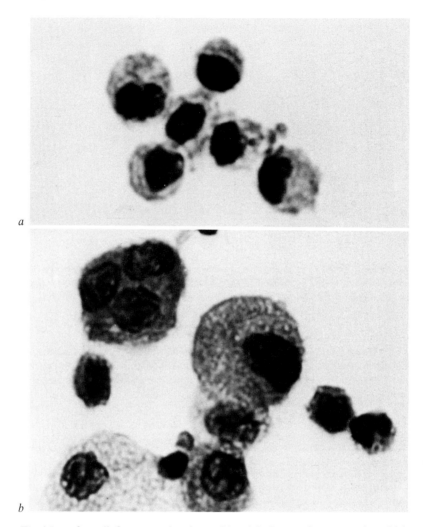

Fig. 4. Lung free cells from control rat lavage (a) and the lavage of rats exposed to a high dose of chrysotile asbestos for 15 weeks (b). Note the variation in cell size and nuclear content of the exposed cells. Glutaraldehyde fix, Giemsa stain ×1,890.

(fibreglass) or a reduction (powdered glass) in the amount of this material isolated (fig. 3).

Lung free cell numbers are shown in combination with the surfactant levels for each experiment (fig. 1–3). The numerical changes in this population did not match those seen with surfactant. High doses of chrysotile

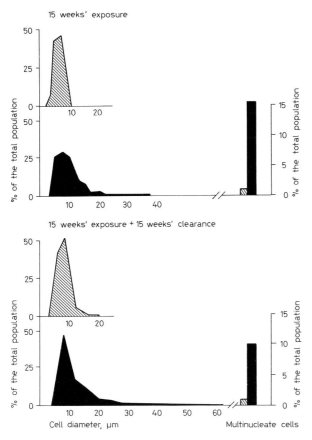

Fig. 5. The effect of exposure to high doses of chrysotile asbestos on rat lung free cell size and the frequency of multinucleate cells. ⊠ = Controls; ■ = test.

resulted in an initial slight drop in lung free cells (fig. 1), but these returned to normal by the end of the 15-week exposure period. After 4 weeks' clearance, the treated cell populations again fell below those of control (0.6 times control level, fig. 2), but rose to above-control levels with 35 weeks' clearance (1.4 times control levels, fig. 1). A similar dose level of amosite asbestos caused no change in lung free cell numbers (fig. 3). Low doses of chrysotile induced a small increase in cell numbers immediately after exposure, which fell below control levels after 12 weeks' clearance (fig. 2). Of the two glass dusts, only the fibreglass sample resulted in increased lung free cells (approximately 1.6 times control levels, fig. 3).

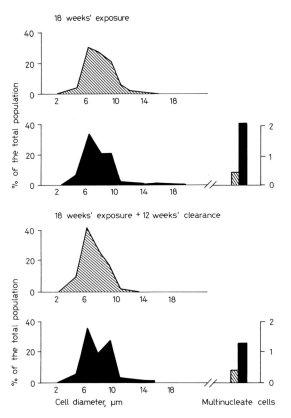

Fig. 6. The effect of exposure to a low dose of chrysotile asbestos on rat lung free cell size and the incidence of multinucleate cells. ▨ = Controls; ■ = test.

Figures 4–6 show chrysotile asbestos-induced changes in free cell morphology. The cells from dust-exposed rats were larger with a higher incidence of multinucleate cells, even after removal from the stimulus (fig. 5) and at low dose levels (fig. 6), although in the latter case this was not so marked. Inhalation of chrysotile and amosite asbestos stimulated the presence of giant or foamy free alveolar cells (fig. 4) while the free cells from amosite-exposed rats often aggregated along the fibres. Although there were a few giant cells following fibreglass exposure [14], the majority of cells, whilst containing particles, was essentially normal in appearance, as were those derived from powder glass-exposed rats.

Discussion

The observations reported suggest that a specifec response occurs at the lung surface following deposition of different minerals. High doses of asbestos promote considerable and possibly permanent alterations in lung surface biochemistry, in particular inducing increased pulmonary surfactant levels and abnormal lung free cell morphology. In direct contrast, inhalation of powdered glass causes reduced surfactant levels with little change in the lung free cell population, while glass fibres primarily affect the lung free cell population, where cell numbers increase and some are morphologically altered.

The turnover of both the lung free cell population and pulmonary surfactant represents a dynamic situation which is dependent on many variables. Thus, increased lung free cells (as seen after fibreglass exposure) could be due to a greater input of these cells to phagocytose the dust, or to a slower rate in their removal from the lung surface. Conversely, the initial drop in the lung free cells following chrysotile exposure may reflect reduced cell output, increased cell removal rates or simply cell death due to the known high cytotoxic nature of the dust. Possibly the cytotoxic potential of a dust is more accurately reflected by changes in free cell morphology. Thus, fibreglass and powdered glass, which rarely induce pathological changes in lung morphology [4–6,11], cause only minor cytological alterations in the free cells. In contrast, inhalation of chrysotile, which induces asbestosis [18], results in a radical change in the lung free cell morphology [15].

In experimentally induced asbestosis the elevation of surfactant levels, which may precede changes in free cell morphology, should also be considered. Some of the observed cytological changes, such as the appearance of foam cells in chrysotile-treated rats, may well be related to excessive phospholipid (and hence surfactant) accumulation at the lung surface, an effect already described by *Lüllman et al.* [9] following chronic drug administration to rats. There is, however, some disagreement as to the role of the free cell in the primary catabolism of surfactant [cf. 3,12] and thus the foam cell may arise from accumulating breakdown products of surfactant or from direct mineral interference with its own lipid metabolism.

Increases in pulmonary surfactant following asbestos inhalation may well be equally or more significant to observed changes in free cell morphology. Elevated lung lipoprotein levels (not specifically pulmonary surfactant) are found in other pathological conditions such as experimental silicosis [7], following use of hypocholesteremic drugs [8] and alveolar lipoproteinosis [13]. From these investigations, theories have been proposed relating to lipoprotein synthesis and degradation under pathological lung conditions. We have suggested that, following chrysotile inhalation, surfactant accumulates due to excessive lipoprotein synthesis, as a result of a numerical increase in type II cells [10] without a corresponding increase in its degradation [16]. However, whilst the route of synthesis of surfactant is generally agreed, the mechanism of catabolism of lung surface lipoproteins is not clearly understood [3].

It can be argued that stimulation of lung surfactant levels may be a primary, non-specific event following lung exposure to toxic materials, relating to changing lung surface tensions or to coat foreign materials and protect lung cells. It is not known, however, whether persistent increases in lung surface lipoprotein will affect subsequent lung function. Does such an event lead to altered lung surface tension or reduced efficiency of gaseous exchange? Are excess levels of surfactant toxic to lung cells; does it prolong alveolar particle retention or mineral/cell interaction by trapping these dusts and hindering lung clearance?

In summary, these studies suggest that initial alterations in pulmonary surfactant levels and lung free cell morphology reflect the nature and dose of the material inhaled and such events may be important in any subsequent pathological changes.

Summary

The effect of inhalation of either asbestos or glass dust on two lung surface components (pulmonary surfactant and lung free cells) was studied. The response monitored depended on both the type of material deposited and the dose level administered. In particular, high doses of asbestos selectively induced increased pulmonary surfactant levels and altered lung free cell morphology, which persisted even after removal from the dust stimulus. Possible reasons for, and consequences of, these results are discussed.

References

1   Desai, R.; Hext, P.M., and Richards, R.J.: The prevention of asbestos-induced haemolysis. Life Sci. *16:* 1931–1938 (1975).
2   Desai, R. and Richards, R.J.: The adsorption of biological macromolecules by mineral dusts. Environ. Res. *16:* 449–464 (1978).

3   Desai, R.; Tetley, T.D.; Curtis, C.G.; Powell, G.M., and Richards, R.J.: Studies on the fate of pulmonary surfactant in the lung. Biochem. J. *176:* 455–462 (1978).

4   Gross, P.; De Treville, R.T.P., and Haller, M.: Pulmonary ferruginous bodies, studies on their origin. Pneumoconiosis, Proc. Int. Congr. Johannesburg 1969, pp. 86–91 (Oxford University Press, London 1970).

5   Gross, P.; Tuma, J., and De Treville, R.T.P.: Lungs of workers exposed to fibreglass. A study of their pathologic changes and their dust content. Archs. envir. Hlth *23:* 67–76 (1971).

6   Gross, P.; Westrick, M.L., and McNerney, J.M.: Glass dust: A study of its biological effects. AMA Archs. Ind. Hlth *21:* 10–23 (1960).

7   Heppleston, A.G.; Fletcher, K., and Wyatt, I.: Changes in the composition of lung lipids and the 'turnover' of dipalmitoyl lecithin in experimental alveolar lipo-proteinosis induced by inhaled quartz. Br. J. exp. Path. *55:* 384–395 (1974).

8   Kikkawa, Y. and Suzuki, K.: Alteration of cellular and acellular alveolar and bronchiolar walls produced by hypocholesteremic drug AY9944. Lab. Invest. *26:* 441–447 (1972).

9   Lüllman, H.; Lüllman-Rauch, R., and Wasserman, O.: Drug induced phospholipidoses. CRC crit. Rev. Toxicol.: 185–218 (1975).

10 McDermott, M.; Wagner, J.C.; Tetley, T.D.; Harwood, J.L., and Richards, R.J.: The effects of inhaled silica and chrysotile on the elastic properties of rat lungs; physiological, physical and biochemical studies of lung surfactant; in, Inhaled particles and vapours, 4th, pp. 415–428 (Pergamon Press, Oxford 1977).

11 Morgan, W.K.C. and Seaton, A.: Occupational lung diseases, p. 116 (Saunders, Philadelphia 1975).

12 Naimark, A.: Cellular dynamics and lipid metabolism in the lung. Fed. Proc. *32:* 1967–1971 (1973).

13 Ramirez-R, J. and Harlan, W.R.: Pulmonary alveolar proteinosis. Am. J. Med. *45:* 502–512 (1968).

14 Tetley, T.D.: Cellular and biochemical effects of asbestos in the rabbit and rat lung. PhD thesis University of Wales, Cardiff (1978).

15 Tetley, T.D.; Hext, P.M., and Richards, R.J.: Chrysotile-induced asbestosis: changes in the free cell population, pulmonary surfactant and whole lung tissue of rats. Br. J. exp. Path. *57:* 505–513 (1976).

16 Tetley, T.D.; Richards, R.J., and Harwood, J.L.: Changes in pulmonary surfactant and phosphatidylcholine metabolism in rats exposed to chrysotile asbestos. Biochem. J. *166:* 323–329 (1977).

17 Timbrell, V.; Skidmore, J.W.; Hyett, A.W., and Wagner, J.C.: Exposure chambers for inhalation experiments with standard reference samples of asbestos of the International Union Against Cancer (UICC). Aerosol Sci. *1:* 215–223 (1970).

18 Wagner, J.C.; Berry, G.; Skidmore, J.W., and Timbrell, V.: The effects of the inhalation of asbestos in rats. Br. J. Cancer *29:* 252–269 (1974).

Dr. T.D. Tetley, Department of Biochemistry, University College, PO Box 78, Cardiff CF1 1XL (Wales)

Prog. Resp. Res., vol. 15, pp. 104–112 (Karger, Basel 1981)

The Effect of Amphiphilic and Antidepressant Drugs on Lung Surface Components Including Surfactant in Experimental Animals[1]

R.J. Richards, R. Lewis and J.L. Harwood

Department of Biochemistry, University College, Cardiff

Introduction

Recently, *Lüllman et al.* [4] have reviewed studies related to the drug-induced phenomenon known as phospholipidosis whereby single, or more usually repeated administrations of amphiphilic compounds to experimental animals results in the accumulation of lipid-like structures in various tissues and cells. One such drug, chlorphentermine, was found to exhibit differential tissue accumulation, mostly as the unchanged drug, following chronic exposure in experimental rats. In particular, the adrenal glands and the lungs had a high tissue/plasma drug level in the chronically exposed animals. As much of the lipid and phospholipid metabolism in the lung is concerned with the formation/degradation of pulmonary surfactant, our original interest was to determine whether amphiphilic compounds would be useful in manipulating surfactant metabolism. When it was evident that such drugs could promote the level of this material isolated from the lung surface, further questions were posed. How do the drugs promote surfactant levels? Is this affect drug-specific and/or are other lung components affected? Is there any clinical significance and/or are there any physiological consequences of such interaction?

The experiments described below represent some initial studies to provide answers to these questions and are particularly concerned with the effects of the tricyclic antidepressant imipramine on the lung surface of experimental rats.

[1] Supported by a grant from the Medical Research Council.

Materials and Methods

Animals and Drug Administration

For intubation experiments, male Sprague-Dawley rats were given chlorphentermine hydrochloride (10 mg/kg body weight/day) or imipramine hydrochloride (75 mg/kg body weight/day) in phosphate-buffered saline for 5 days a week over a period of 2 weeks. In the first experiment both drugs were used, whereas imipramine alone was used in the second experiment. Control animals were given phosphate-buffered saline only.

For the feeding studies, male MRC hooded rats (departmental strain), approximately 4 months old, were fed with Tofranil (imipramine hydrochloride) in a crushed tablet form which was mixed with approximately 1.5 g margerine. The doses employed were 1, 5, 10 and 50 mg/kg body weight/day for periods up to 8 weeks. Control animals received margerine only and all animals were given the normal rat pellet diet and water *ad libitum.* The highest dose (50 mg) was unacceptable to the rats after 3 days and attempts to replace the margerine/drug diet with the drug in syrup form also proved unsuccessful.

For the intraperitoneal injection studies, female MRC hooded rats, 4 months old, were given a daily injection of Tofranil (intramuscular injection form) at 32 mg/kg body weight/day for 6 consecutive days, the drug being diluted to the required dosage with 0.15 M saline. The injected volume was 0.5 ml and control animals received 0.5 ml of saline only. The animals were sacrificed at 6, 12 and 18 days after the first injection.

For the intratracheal instillation studies, female MRC hooded rats, 4 months old, were anaesthetised with halothane and given a single instillation (0.25 ml) of the intramuscular form of imipramine hydrochloride which had been suitably diluted in sterile 0.15 M saline. The doses chosen were 0.03, 0.06 and 0.12 mg/rat (0.15, 0.30 and 0.60 mg/kg body weight, respectively) and control animals received 0.25 ml of saline alone. The rats were sacrificed at 2, 4 and 8 days after the single instillation.

Treatment of Lung Tissue

All animals were sacrificed by CO_2 asphyxiation, the lung circulation perfused with saline and the lungs removed and lavaged with 0.15 ml of saline (6 × 10 ml washes). The washes from each animal were pooled, a cell count performed and a smear made for subsequent cytological examination. The lavaged lung was dried between filter paper, weighed and stored at −20°C for future analysis. In some experiments the liver and brain were also removed and stored at −20°C. With the exception of the feeding experiments the lavage fluid was processed further. In the intraperitoneal injection studies the free cells were separated from the remainder of the lavage, in the intubation study the free cells and pulmonary surfactant were isolated and in the instillation experiments the free cells, pulmonary surfactant and a fraction containing the bulk of soluble lung protein were isolated. Details of these procedures are given below.

Fractionation of Lung Lavage Fluid.

The free cells were removed from the lavage fluid by centrifugation at 300 g for 20 min. The pellet was washed with 0.15 M saline and the mixture recentrifuged (300 g for 20 min) after which the supernatant is added to the supernatant from the first spin. The pellet constitutes the washed free cell population and the combined supernatants can be examined for lipid/protein components (as with the intraperitoneal injection studies) or further processed

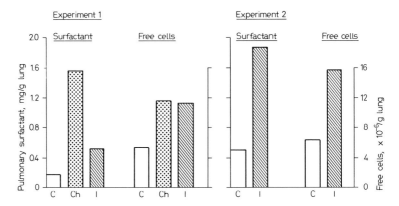

Fig. 1. Changes in free cell numbers and pulmonary surfactant derived from rats receiving chlorphentermine (Ch) or imipramine (I) by stomach intubation. C = Controls.

to obtain specific fractions. The supernatant was centrifuged at 1,000 *g* for 60 min and the resulting clear supernatant contains the majority of alveolar surface protein [1]. The small pellet was resuspended in 21% saline, mixed well and centrifuged at 1,500 *g* for 25 min whereupon a three-phase separation was achieved. The top layer, or pellicle, was designated pulmonary surfactant and was carefully removed and dialysed against distilled water for 24 h. This material was then freeze-dried and weighed. In order to achieve accurate quantitative results the pooled lavages of at least 3 rats (preferably 6) should be used to isolate the surfactant fraction. The pulmonary surfactant isolated by this method has a lipid/protein ratio of 9 : 1 and when analysed dipalmitoyl phosphatidylcholine accounts for approximately 70% of the lipid moieties [2].

Biochemical Analyses

The extraction, separation and estimation of lipids and total fatty acids in the fractionated samples was carried out as previously described [2]. Estimations of protein were also carried out as described previously [1].

Results and Discussion

Intubation Studies

Chlorphentermine and imipramine both initiate accumulation of pulmonary surfactant and a numerical increase in the free cells 2 weeks after chronic administration by intubation (fig. 1). Many of the free cells are large and have the typical foamy cell appearance as previously described by *Karabelnik et al.* [3]. The effect of chlorphentermine is further emphasised by a comparison of the freeze-dried surfactant fraction obtained from 3 drug-treated rats with that derived from 6 control (normal)

Fig. 2. Freeze-dried preparations of pulmonary surfactant derived from 6 control animals and 3 rats treated with chlorphentermine by intubation.

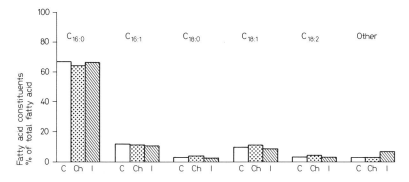

Fig. 3. Fatty acid composition of isolated surfactants derived from control (C) chlorphentermine (Ch) and imipramine-treated rats (I) (intubation study).

animals (fig. 2). Analysis of the fatty acid composition of the surfactant fractions from each animal group suggests there is little difference in lipid composition of the material derived from normal and drug-treated rats (fig. 3). Separation of the acyl lipid components and quantitation of their fatty acid constituents suggested that the drug-induced increase in surfactant did not result in material of abnormal composition being produced. It should, however, be mentioned that the isolated surfactant represents only one fraction of the lavage fluid and changes in lipid/protein components in the other fractions cannot be discounted. From these experiments it can be concluded that sufficient quantities fo the orally administered drug or a metabolite therefrom can reach the lung and stimulate surfactant accumulation. *Lüllman et al.* [4] have shown that chlorphentermine accumulates unchanged in lung (and other) tissues which

Table 1. Changes in tissue and tissue fractions isolated from rats given intraperitoneal injections of imipramine (32 mg/kg body weight/day) for 6 days followed by cessation of treatment for 6(12-day samples) and 12 days (18-day samples), respectively.

Treatment	Days after first injection	Number of free cells ×10^{-6}/g lung	Protein in lavage fluid mg/g lung	Total lipid in lavage fluid mg/g lung	Protein in lung tissue mg/g lung	Lipid in lung tissue mg/g lung	Liver protein mg/g lung	Brain protein mg/g	Changes in fatty acid composition lavage fluid	Changes in fatty acid composition lung tissue
Control	6	7.95 (5.40–9.46)	5.06 (4.36–5.76)	4.47 (4.04–4.90)	58.3 (52.5–66.2)	23.6 (19.8–29.9)	78.7 (65.0–87.5)	65.0 (55.0–73.8)		
Imipramine	6	10.85 (9.32–12.10)	7.50 (5.72–9.50)	6.07 (3.44–10.67)	61.8 (56.8–65.0)	21.2 (14.8–32.8)	79.2 (77.5–80.0)	68.3 (67.5–70.0)	$C_{16.0}$↑	$C_{14.0}$↓ $C_{18.0}$↑
Control	12	12.28 (9.83–16.72)	5.40 (4.92–5.76)	4.50 (4.10–4.72)	81.3 (68.8–93.8)	38.8 (33.0–44.5)	81.7 (76.2–86.2)	57.6 (48.8–62.5)		
Imipramine	12	9.50 (6.78–10.97)	7.07 (5.30–11.22)	4.55 (3.44–5.70)	68.6 (61.3–72.0)	20.4 (16.6–23.1)	86.7 (83.7–91.2)	62.5 (53.7–68.8)	–	$C_{18.2}$↓
Control	18	8.52 (6.67–10.21)	5.20 (4.50–5.84)	4.63 (3.23–6.02)	57.9 (55.0–60.0)	22.0 (17.1–27.0)	83.4 (80.6–87.5)	55.5 (52.5–56.3)		
Imipramine	18	10.18 (8.16–12.38)	6.22 (5.76–6.85)	5.10 (4.91–5.34)	57.9 (47.5–63.8)	31.8 (23.1–43.6)	76.1 (63.2–96.2)	53.4 (48.7–57.5)	–	$C_{14.0}$↓ $C_{16.0}$↑ $C_{18.0}$↓

Ranges given in parentheses. The lavage fluid is that fraction which does not include free cells. ↑ = Significantly higher than control; ↓ = significantly lower than control (Students' t test). Results are from 3 rats in each group.

suggests that the intact drug and not a metabolite may directly mediate surfactant accumulation. However, other factors relating to the toxicity of the drug in the lung (see instillation studies) may be important in this effect.

Feeding Studies

In these preliminary experiments a cytological examination only was made of the free cells in the lung lavage material. Few, if any, foamy cells were observed in the lavage from animals ingesting 1, 5 or 10 mg imipramine/kg body weight/day over a period of 8 weeks. These observations may be taken as an indication of the absence of any extensive phospholipidosis reaction. However, no quantitative biochemical estimations for surfactant, total lavage lipid or protein were made and therefore a minimal phospholipidosis reaction cannot be excluded. The only tentative conclusion drawn from this study is that natural ingestion of at least 10 mg of drug/kg body weight/day (for comparative purposes: normal human intake of imipramine is approximately 1 mg/kg body weight/day) for 8 weeks does not result in sufficient intact drug reaching the lung tissue to initiate a phospholipidosis reaction. Attempts to increase the animal intake of the drug were mostly unsuccessful and when rats ingested 50 mg drug/kg body weight/day the effect was sometimes fatal (75% survival rate over 2 weeks).

Intraperitoneal Injection Studies

Chronic intraperitoneal administration of imipramine produces little alteration in free cell numbers (table I), although small numbers of foamy cells were present in the lavage from some rats, thus suggesting a weak phospholipidosis reaction. Only 75% of animals survived the injection period of 6 consecutive days and inflammatory lesions were noted at the injection sites of drug-treated animals and in most instances the liver was considerably reduced in size at 6 days. The latter effect was quickly rectified upon cessation of drug treatment. Changes in the mean lavage total protein and lipid upon completion of treatment (6 days) indicates that the drug promotes accumulation of both these components (table I). The range values show that some animals are affected considerably more than others by the drug despite the equivalent treatment for each rat. After cessation of treatment lavage lipid levels are similar to those found in control rats although some drug-tested animals still retain elevated protein in the lavage fluid. Some significant differences in fatty acid composition are detected in the lavage fluid and remaining lung tissue of control rats and those treated with imipramine up to 12 days after cessation of treatment, but these were not extensive. Any changes in pulmonary surfactant were not assessed in these experiments, although elevation of this component is usually indicated by an increase in alveolar surface lipid. As was found with the feeding studies it is possible that insufficient quantities of intact drug reached the lung in the time period of experimentation to produce such extensive changes as those seen with the intubation study. Few changes were detectable in total lung tissue lipid or protein or in the brain and liver protein levels of drug-treated animals (table I).

Intratracheal Instillation Studies

These investigations were designed to extend the observations from the previous two studies in that possibly imipramine would be reactive upon making direct contact with lung tissue. Preliminary experiments indicated that direct intratracheal instillation of a single dose of imipramine in excess of 1 mg/kg body weight was usually fatal. Single doses of imipramine of 0.6, 0.3 and 0.15 mg/kg body weight were therefore chosen for the experiments. 2 days after the drug instillation little, if any, changes were detected in the free cell numbers, lavaged protein was lower with higher concentrations of drug and the amount of pulmonary surfactant was slightly elevated in the lowest group of drug-treated rats when compared with values found in normal animals (fig. 4). 4 days after drug instillation all doses of imipramine cause a considerable elevation in isolated surfactant in the absence of any significant changes in free cell numbers, although the protein in the lavage fluid is higher than that found in control rats. 8 days after drug instillation lavage protein levels are directly related to dose of instilled drug, the highest dose (0.12 mg) causing a 4-fold increase in this component. At this time period the number of free cells decreases with increasing concentration of instilled imipramine (fig. 4). Elevated levels of surfactant are also maintained 8 days after the drug instillation, although there is a gradual decline in the amount of surfactant with increasing concentration of imipramine.

The results indicate that imipramine exerts a time- and dose-dependent effect at the lung surface. It is difficult to visualise, however, the precise events which accompany the drug-induced accumulation of surfactant. In most instances the lowest dose of drug is most effective in maintaining a higher level of surfactant. There are many possible explanations for this effect, but the most likely is that higher concentrations of drug are more toxic to lung cells. Evidence for this comes from the reduc-

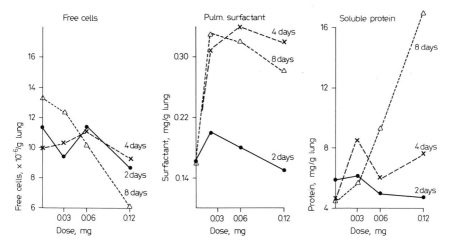

Fig. 4. Changes in free cell numbers, pulmonary surfactant and lavaged soluble protein following different doses of intratracheally instilled imipramine over a time period of 8 days. Values for control animals are shown on the vertical axis.

tion in free cell numbers and elevated surface protein particularly after 8 days. Elevation in protein levels may result from interstitial leakage of plasma proteins and/or accumulation of protein from dead and damaged cells. In this regard any newly formed or active type II cells may also be damaged by increasing drug concentration and thus account for the surfactant/drug concentration curves shown in figure 4.

In summary, in this experiment surfactant accumulation may be a secondary response to the primary cellular damage at the lung surface induced by imipramine. Whether or not a subtoxic dose of this particular drug can induce a direct, specific effect on surfactant synthesis requires further investigation.

Summary

Chlorphentermine and imipramine, chronically adminstered to rats by intubation, can both initiate the accumulation of a lipoprotein-rich fraction, isolated from the lung surface and designated pulmonary surfactant. Such a reaction may be related to the toxicity of the drug following its accumulation in lung tissue. This suggestion is supported by the fact that small quantities of imipramine directly instilled intratracheally promote dose- and time-dependent changes in the free cell population, alveolar surface-soluble protein and pulmonary surfactant. Whilst this experimental model system may be useful to compare the relative toxicities of drugs likely to reach lung tissue and provide a means of manipulating surfactant metabolism, the clinical manifestations and physiological consequences of drug-induced surfactant accumulation require further investigation.

References

1   George, G. and Richards, R.J.: Preliminary studies on the isolation, separation and iden-
    tification of pulmonary lavage proteins from the rabbit. Biochem. Soc. Trans. *7:*
    1285–1287 (1979).
2   Harwood, J.L.; Desai, R.; Hext, P.M.; Tetley, T.D., and Richards, R.J.: Characterisa-
    tion of pulmonary surfactant from ox, rabbit, rat and sheep. Biochem. J. *151:* 707–714
    (1975).
3   Karabelnik, D.; Zbinden, G., and Baumgartner, E.: Drug-induced foam cell reactions
    in rats. 1. Histopathologic and cytochemical observations after treatment with chlor-
    phentermine RM1-10393 and R04-4318. Toxicol. appl. Pharmacol. *27:* 395–407
    (1974).
4   Lüllman, H.; Lüllman-Rauch, R., and Wasserman, O.: Drug-induced phospholipi-
    doses. CRC crit. Rev. Toxicol. *4:* 185–218 (1975).

Dr. R.J. Richards, Department of Biochemistry, University College, PO Box 78, Cardiff
CF1 1XL (Wales)

Discussion

Von Wichert commented on the experiments of *Richards,* referring to some experiments with amphiphilic drugs in the sense of *Lüllmann.* Using bromhexine, he showed an increase in phospholipid content in animals and he asked *Richards* if this could be induced by an alter-ation in the balance between free phospholipid molecules on the cells and those phospholipid molecules which are captured by the amphiphilic drug. Further it has been shown that the increased amount of phospholipids during experiments with chlorphentermine leads to an increase in surfactant activity. *Richards* answered, there may be a compositional change in pulmonary surfactant under the influence of amphiphilic drugs, but these changes are rather small. He had not observed such changes. *Clements* asked *Richards* and *Tetley* to speculate on the role of lymphocytes in alveolar clearance because it has been shown that they are immunologically suppressed by surfactant material. *Richards* answered that he had never found lymphocytes in the lavage even under quite serious pathological conditions, he had, however, found neutrophilic leucocytes. *Burkhardt* asked *Tetley* about the interpretation of foamy cells as a sign of cellular degeneration. *Tetley* had measured a lot of enzyme activity in these cells. The foamy cells may take in the lipid of the lung surface but not necessarily break down these lipids. On a second question by *Burkhardt, Tetley* said that there is an increase in type II cells following chrysotile deposition. *Reiss* added the comment that amphiphilic substances not only act on the lipids but also on the proteins, because they may also be hydro-phobic and it is an incorrect assumption that these drugs act on the lipids, they could very well act on the proteins.

Prog. Resp. Res., vol. 15, pp. 113–120 (Karger, Basel 1981)

Changes of the Lecithin Content in the Rat Lung under the Influence of Toxic Doses of Hypnotics and Nonphysiological Oxygen Concentrations

D. Barckow, G. Kynast and T. Schirop

Resuscitation Center of the Medical Clinic and Outpatient Department of the Free University of Berlin at the Charlottenburg Clinic (Department Director: Prof. Dr. *K. Ibe)*, Berlin

Introduction

Today, the clinical and experimental interests of many investigators are concentrated on the surfactant-producing system of the lung and the factors which may disturb this system. This is understandable as a disturbance of the antiatelectase factor plays an important role in the early phase of the 'acute respiratory distress syndrom', no matter if it occurs in newborns or in adults [3, 8, 15, 19, 20].

In the adult especially, quite a number of toxic agents which are thought to influence the normal production and composition of surface-active substances of the lung are discussed [2, 7, 11, 13]. Also, hypnotics may possibly have to be classified with these toxic agents, since it is not unusual that an 'acute respiratory distress syndrome' occurs in connection with severe hypnotic intoxications. This can become a real problem even for the modern intensive care unit [18]. In animal experiments, therefore, we have tried to find out to what extent the hypnotic drugs bromcarbamide and phenobarbital, which play an important role in human toxicology, have the presumed side effect on the production or composition of the lung surfactant as represented by lecithin. Another point of interest was to find out whether the supposed damaging influence of both of these substances can become more aggravated by nonphysiological oxygen concentrations; quite often necessary for therapeutical reasons in cases of poisoning in humans.

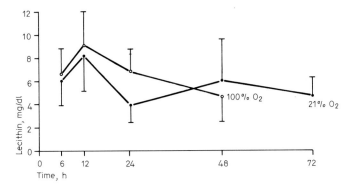

Fig. 1. Lecithin content of lung lavage fluid in rats after bromcarbamide intoxication and while breathing air or pure oxygen.

Material and Methods

The experiments were conducted on a total of 350 Wistar rats of both sexes, weighing between 180 and 220 g. Groups of 10 animals were sub-lethally intoxicated with bromcarbamide (200 mg/kg body weight) or phenobarbital (100 mg/kg body weight) and were exposed to room air or to an atmosphere of pure oxygen during the period of observation.

Because of their poor solubility in water, both hypnotics, after solution in olive oil, were administered to the animals by intraperitoneal injection. Earlier investigations with regard to the toxicology and pharmacokinetics of the substances named were carried out by us in a similar manner.

The duration of coma of the intoxicated animals lasted for between 8 and 12 h according to the dosage chosen.

After 6, 12, 24, 48 and 72 h, one group of animals was killed. The isolated lungs were inflated over a tracheal tube and subsequently rinsed with 15 ml of physiologic saline. 2 ml of this rinsing solution were extracted immediately afterwards with methanol and chloroform in order to determine the lecithin content. This determination was established by thin-layer chromatography [10].

Up to now, in 50 animals the remainder of the lung lavage fluid was used to determine the surface activity in the Wilhelmy balance [6, 8, 12, 16]. Control animals received an oil injection only.

Results

With the chosen method of determination, the lecithin content in the lung lavage fluid of untreated animals was at 3.5 ± 1.2 mg% and this was within the limits found by other investigators [5]. Under the influence of

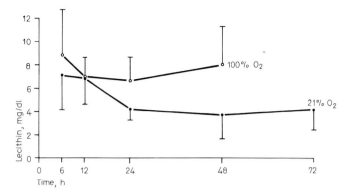

Fig. 2. Lecithin content of lung lavage fluid in rats after barbiturate intoxication and while breathing air or pure oxygen.

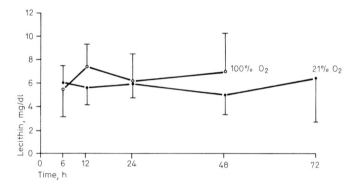

Fig. 3. Lecithin content of lung lavage fluid in rats after intraperitoneal application of olive oil alone and while breathing air or pure oxygen.

bromcarbamide intoxication (fig. 1) this value considerably increased directly after administering the poison, and reached its maximum of 8.1 mg% after about 12 then slowly returned to normal.

Pure oxygen further intensified this effect even more. The intoxication with phenobarbital (fig. 2) had a similar effect, but here the final extent of the changes was measurable earlier. In addition, the lecithin content of the lung lavage fluid, under the added pure oxygen, remained considerably elevated during the whole time of observation. Olive oil alone likewise

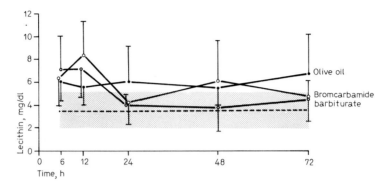

Fig. 4. Lecithin content of lung lavage fluid in rats after intoxication with barbiturates or bromcarbamides while breathing air. Shaded area indicates normal values.

increased the lecithin content (fig. 3), but the changes were much less distinct.

It is to be noted that none of the animals survived their confinement in pure oxygen for more than 56 h, whether intoxicated or not.

Discussion

The established results raise a number of questions. Quite obviously (fig. 4), it was shown that the rats examined by us, whilst in a coma due to the two hypnotics, developed a considerable increase of lecithin content in the lung lavage fluid. Breathing room air, this rise quickly faded after the symptoms of poisoning had disappeared. By adding pure oxygen (fig. 5) this effect was substantially intensified and the lecithin values remained remarkably high during the entire period of observation. However, compared with results obtained from animals which were exposed to pure oxygen only, the difference was very slight. After a 48-hour test period, electron microscopy (fig. 6) showed that the alveolar type II cells from the rats were almost completely degranulated.

In spite of that, especially in these animals, the lung function at that time was significantly reduced. This could be shown by the distinct decrease in lung compliance which we examined in all cases [4].

From this it may be concluded that the lungs of rats intoxicated by us contained sufficient surface-active material, but perhaps this could not

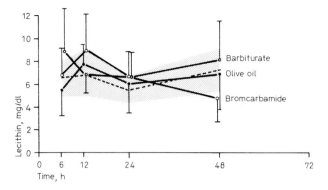

Fig. 5. Lecithin content of lung lavage fluid in rats after intoxication with barbiturates or bromcarbamides while breathing pure oxygen. Shaded area indicates control group.

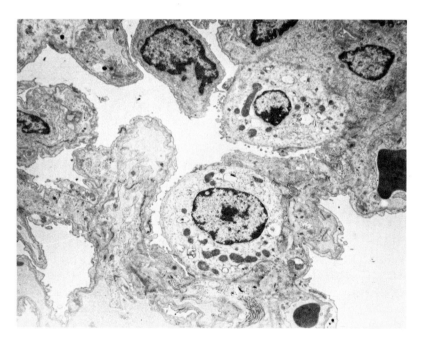

Fig. 6. Degranulation of alveolar type II cell from the rat 48 h after intoxication and breathing pure oxygen.

develop its specific effect on the alveolar micromechanics because it was detached from the alveolar surface. The strongly reduced surface activity in many of the samples from intoxicated animals as obtained by the Wilhelmy balance [6] also allows the conclusion that the surfactant was reduced in quality by the noxious matter chosen. Only further tests can show the precise mechanisms of these disturbances.

For practical purposes, it now seems important to know about the speed [9] at which the changes described here develop, even though this is, only at present, found in animal experiments. In addition, lung poisoning due to nonphysiological oxygen concentrations is documented again [7, 14]. By applying these results to humas we may perhaps obtain a certain hint as to why early artificial ventilation with positive end-expiratory pressures has proved to be a rather effective therapy [1, 17, 21], for no other kind of ventilation is more able to compensate an irritation of the lung surfactant.

Summary

350 Wistar rats were sub-lethally intoxicated with bromcarbamide or phenobarbital. At various observation periods we examined, if the drugs changed the lecithin content of the lung lavage fluid in these animals. Lecithin, which is regarded as being the most important component part of the lung surfactant, was determined by thin-layer chromatography. Both hypnotics increase the lecithin content by double the normal value, reaching a maximum at about 12 after administering the drugs.

The additional breathing of pure oxygen slightly increases this effect. Inspite of this lung compliance, expecially in those animals which did not survive more than 56 h, had changed significantly. From these results we conclude that the surface-active system of the rat lung under the influence of intoxication, and especially by adding pure oxygen, is either damaged in its quality or is separated from the alveolar surface, thus losing its normal function.

References

1   Barckow, D.; Schirop, T.; Zimmermann, D.; Loddenkemper, R.; Ohlmeier, H., und Korsukewitz, J.: Möglichkeiten zur Beeinflussung pulmonaler Komplikationen bei schweren Schlaftmittelvergiftungen. Akt. Probl. Intensivmed. *2:* 204–209 (1976).

2   Baum, M.; Benzer, H.; Blümel, G.; Bolcic, J.; Irsigler, K. und Tölle, W.: Die Bedeutung der Oberflächenspannung in der Lunge beim experimentellen posttraumatischen Syndrom. Z. exp. Chirurg. *4:* 359–376 (1971).

3   Benzer, H.: Respiratorbeatmung und Oberflächenspannung in der Lunge. Anaesthesiology and resuscitation, Vol. 38 (Springer, Berlin 1969).

4   Bone, R.C.: Diagnosis of causes for acute respiratory distress by pressure-volume curves. Chest *70:* 740–746 (1976).

5   Brody, A.R.; Clay, M.F.; Collins, M.M.; Eden, A.C., and McDermott, M.: Effects of chlorphentermine hydrochloride on the surface tension properties of the rat lung. J. Physiol. *245:* 105–106 (1975).

6   Clements, J.A.; Hustead, R.F.; Joanson, R.P., and Gribetz, I.: Pulmonary surface tension and alveolar stability. J. appl. Physiol. *16:* 444–450 (1961).

7   Gilder, H. and McSherry, C.K.: Phosphatidylcholine synthesis and pulmonary oxygen toxicity. Biochim. biophys. Acta *441:* 48–56 (1976).

8   Goerke, J.: Lung surfactant. Biochim. biophys. Acta *344:* 241–261 (1974).

9   Haider, W.; Baum, M.; Benzer, H. und Lackner, F.: Ablauf der Lungenveränderungen im posttraumatischen Schock (Schocklunge). Anaesthesist *23:* 129–136 (1974).

10   Kynast, G. and Saling, E.Z.: Rapid specific determination of amniotic fluid lecithins as a test of fetal lung maturity. J. perinat. Med. *1:* 213–218 (1973).

11   Landauer, B.; Tölle, W. und Knittel, E.: Zur Beeinflussung der Funktionscharakteristik des Antiatelektasefaktors durch Inhalationsanästhetika. Medsche Welt *28:* 915–919 (1977).

12   Landauer, B.; Tölle, W. und Blümel, G.: Experimentelle Nachweismöglichkeiten der Surfactantaktivität. Medsche Welt *27:* 239–243 (1976).

13   McClenahan, J.B. and Urtnowski, A.: Effect of ventilation on surfactant, and its turn over rate. J. appl. Physiol. *23:* 215–220 (1976).

14   Morgan, T.E.; Finley, T.N.; Huber, G.L., and Fialkow, H.: Alterations in pulmonary surface active lipids during exposure to increased oxygen tension. J. clin. Invest. *44:* 1737–1744 (1965).

15   Renovanz, H.-D. and v. Seefeld, H.: Surfactant and respiratory distress syndrome. Prax. Pneumol. *32:* 443–466 (1978).

16   Rüfer, R.: Experimentelle Grundlagen der Oberflächenaktivität in der Säugetierlunge. Pneumologie *144:* 167–175 (1971).

17   Schmidt, G.B.; O'Neill, W.W.; Kotb, K.; Ko Hnang, K.; Bennett, E.J., and Bombeck, C.T.: Continuous positive airway pressure in the prophylaxis of the adult respiratory distress syndrome. Langenbecks Arch. Chir., suppl. Chir. Forum *1975:* 439–442.

18   v. Wichert, P.; Schmidt, C.; Pomränke, K., and Wiegers, U.: Incorporation of radioactive labelled choline and palmitate into lung lecithin of rabbits treated with high doses of bromcarbamides, barbiturates and diazepam. Arch. Tox. *37:* 117–122 (1977).

19   v. Wichert, P. and Kohl, F.V.: Decreased Dipalmithoyllecithin content found in lung specimens from patients with so-called shock-lung. Intens. Care Med. *3:* 27–30 (1977).

20   v. Wichert, P.: Beziehungen zwischen pathologischen Veränderungen und Phospholipidgehalt der menschlichen Lunge. Pneumonologie *144:* 201–205 (1971).

21   Wyszogrodoski, I.; Kyel-Aboagye, K.; Taeusch, W., Jr., and Avery, M.E.: Surfactant inactivation by hyperventilation: conservation by end-expiratory pressure. J. appl. Physiol. *38:* 461–466 (1975).

D. Barckow, MD, Resuscitation Center of the Medical Clinic and Outpatient, Department of the Free University of Berlin, at the Charlottenburg Clinic, D-1000 Berlin.

Discussion

Wolf emphasized that the pathomechanism in bromcarbamide intoxication is quite complicated, and includes widening of microvessels and changes of the peripheral and pulmonary vascular resistance. *Geiger* asked if the increase of the lecithin content is a result of a breakdown of the type II cells, thereby releasing the intracellular surfactant into the alveoli, but, as *Barckow* answered, the electron microscopic investigation showed that the alveolar type II cells were not dead at the end of the experiments, but completely exhausted of their lamellar bodies.

Prog. Resp. Res., vol. 15, pp. 121–125 (Karger, Basel 1981)

Electron Microscopic Findings in Rats after Inhalation of Detergents

W. Kissler, K. Morgenroth and W. Weller

Institute of Pathology, University of Bochum and Silicosis Research Institute of Bergbau-Berufsgenossenschaft, Bochum

Introduction

Detergents are increasing in importance in coalmining [1,2,4,5]. They are used to reduce the surface tension of the water in water infusion of coal and in water jet blasting. The coal particles are made heavier by the resulting better moistening and fall to the ground more rapidly. In this way, the miners inhale less coal quartz particles, but on the other hand breathe in traces of the detergent.

In view of the possibility of long exposure times, it appears necessary to investigate the action of these wetting agents on the pulmonary parenchyma.

Material and Methods

We carried out an inhalation test on 48 SPF-Sprague-Dawley rats with the detergent Lassageene Z 100 (Hauhinco, Saarbrücken), FRG which is also used in coalmining. The animals were divided into two groups. One group of animals was exposed for 4 h a day to a wetting agent concentration of 6.7 mg/m³ in the air they breathed and the second group was exposed for 1 h a day to a wetting agent concentration of 22.8 mg/m³. Animals from each group were killed together with control animals after 4, 6, 8 and 10 weeks exposure time. The animals were killed during Nembutal anesthesia after a 72-hour, exposure-free interval. Their lungs were processed for light and electron microscopy after perfusion fixation with 3% glutaraldehyde.

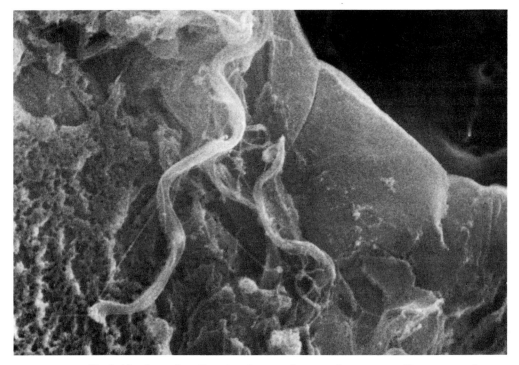

Fig. 1. Alveolar surface, Scanning electron microgram. In some cases filamentous and in other cases amorphously detached components of the surface film adhere to the alveolar wall.

Results

Pathological changes which could be recognized by light microscopy were present in group I (4 h exposure to a low concentration of Lassageene Z 100) after 6 weeks of exposure, and in group II after 8 weeks (exposure to high Lassageene Z 100 concentration for 1 h). The findings were qualitatively the same in the two groups. A further progression of the alterations could also not be demonstrated after a longer exposure time.

Especially noticeable were focal disturbances of ventilation in the lungs extending to complete alveolar collapse. In contrast to other authors [6], we were unable to demonstrate any increased alveolar macrophages in lung areas which were underventilated due to the wetting agent.

In these underventilated regions of the lung, pronounced damage to the surface film could be observed under the electron microscope. Focally, a large vesicular detachment of the osmiophilic surface layer became

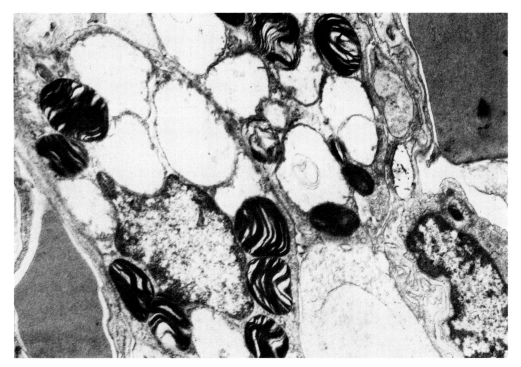

Fig. 2. Greatly damaged type II pneumocyte, pronounced hydropic swelling of the mitochondria, in some cases densely arranged osmiophilic lamellar bodies.

visible. This was evidently based on penetration of water into the basal
layer of the surface film. In scanning electron microscopy, these detach-
ments were seen as filamentous superimpositions on the rest of the intact
surface film (fig. 1).

However, not only was the surface film partially destroyed in the
dystelectatic lung regions, but also the cells which form the surfactant, i.e.,
the type II pneumocytes. Besides a discrete invasion of water into the cytoplasmic matrix, these showed an extreme vacuolar transformation of the
mitochondria. The mitochondrial matrix, which normally consists of fine
granules, showed a high degree of disintegration. The mitochondrial
cristae could only be distinguished as isolated stump-like elevations on the
inner membrane. At the same time, a collection of densely arranged osmiophilic lamellar bodies developed in the cytoplasm of the type II pneumocytes (fig. 2).

In a few pneumocytes, a flask-like apical distension of the cell had developed showing a fine granular matrix without cellular organelles. In areas with total alveolar collapse, complete necrosis of type II pneumocytes could also be observed. The cell membranes disrupted and the constituents of the cytoplasm were released into the alveolar space.

Similar severe degenerative alterations as in type II pneumocytes could be observed in the Clara cells of the bronchioli. These also showed vacuolar degenerations of their mitochondria. Moreover, flask-like apical cell distensions were seen again. In some cases, these filled the lumen of the bronchioli. This finding was especially evident in scanning electron microscopy. The ciliated cells, on the other hand, showed a lower degree of damage than the Clara cells. The cilia were unchanged in arrangement and structure.

Discussion

The findings described, especially the large degree of destruction of the surfactant and numerous type II pneumocytes, explain the occurrence of dystelectasia. The cause of the surfactant damage is probably the good solubility of the phospholipids, the main constituent of the surfactant, in wetting agents.

The disruption of the membranes of the type II pneumocytes and the Clara cells is probably induced by a selective dissolving out of membrane lipids from the cell walls, as has been proved for the lysis of erythrocytes by wetting agents [3]. On the other hand, the relative resistance of the ciliated cells of the bronchioli to the wetting agent Lassageene Z 100 is surprisingly maintained.

Although the pathological alterations of the lungs regress after a long period free of exposure, an application of wetting agents in coalmining does not, according to our results, appear to be unproblematical, since the surfactant of the lungs in particular is destroyed by such agents. The concentration of wetting agent (6.7 mg/m³ respiratory air) which we used is only 100 times higher than the concentrations of wetting agent which are in fact used in coalmining.

Summary

We carried out an inhalation test on 48 SPF–Sprague-Dawley rats with the detergent Lassageene Z 100, which is also used in coalmining. In the lungs of the animals a pronounced damage to the surfactant could be observed under the electron microscope. The type II pneumocytes and the Clara cells showed an extreme vacuolar transformation of the mitochondria. Later, the cell membranes of the pneumocytes disrupted. The alteration of the lung proved to be reversible after a period of nonexposure.

References

1   Bauer, H.D.: Tränken mit Calciumchloridzusatz. Eine Möglichkeit zur Verbesserung der Tränkwirkung. Bergfreiheit *35:* 26–33 (1970).
2   Bauer, H.D.: Möglichkeiten zur Verbesserung der Wirksamkeit von Staubbekämpfungs-verfahren. Glückauf *106:* 309–321 (1970).
3   Cooney, D.A. and Drake, J.C.: The stromatolysis of sheep erythrocytes by Triton X-100. Expl. Cell. Res. *54:* 11–16 (1969).
4   Klinker, H.G. und Bauer, H.D.: Versuche zur Verbesserung der Staubbekämpfung beim Blasversatz, insbesondere durch Verwendung grenzflächenaktiver Stoffe. Bergfreiheit *34:* 198–202 (1969).
5   Seewald, H.; Klein, J.; Breuer, H. und Jüntgen, H.: Untersuchungen über die Benet-zungseigenschaften von Stäuben und über die optimale Anwendung geeigneter Zusatz-mittel für die Verbesserung der Staubbekämpfung mit Wasser; in Ergebnisse von Unter-suchungen auf dem Gebiet der Staub- und Silikosebekämpfung im Steinkohlenbergbau, vol. *11;* pp. 47–61 (Verlag Glückauf, Essen 1977).
6   Zajusz, K.; Nowak, B.; Zylka, M., and Byczkowska, Z.: Zmiany w pluchach szczutow wywolane inhalacjami detergentow. Bromat. chem. toksykol. *10:* 303–309 (1977).

PD. Dr. med. W. Kissler, Institut für Pathologie der Ruhr-Universität, Universitäts-strasse 150, D-4630 Bochum (FRG)

Discussion

Answering a question of *Clements, Kissler* said that the detergent used is a nonionic tensid. Concerning the question of *Bleifeld* on the time course of the alterations, *Kissler* pointed out that the first alteration could be recognized after 6 weeks exposition. *Burkhardt* asked about the consequences of long-term chronic exposure and about symptoms related to this exposure in man and *Kissler's* answer was that the concentration used in the experiments is 100 times of that used in coalmining.

Prog. Resp. Res., vol. 15, pp. 126–132 (Karger, Basel 1981)

Correlation of the Procoagulant Activity in Amniotic Fluid to Gestational Age and Fetal Pulmonary Maturity

W. Leucht, H. Heyes and K. Musch

Department of Obstetrics and Gynaecology of the University of Ulm, Ulm

Introduction

According to the basic research of *Clements* [3], *Avery and Mead* [1], and *Gluck et al.* [5], the fetal lung maturity depends on a sufficient amount of phospholipids in the amniotic fluid (AF). These phospholipids are an essential part of the surfactant factor [3] which protects the lung against the respiratory distress syndrome (RDS). Proportional to the gestational age, there is an increasing procoagulant activity in the AF [13]. This activity was characterized as an activating principle of coagulant factor X acting in the same manner as Russel's viper venom [10]. It is the purpose of this paper to examine whether there is a correlation between the procoagulant activity and the fetal lung maturity.

Material and Methods

AF Sampling

235 amniotic fluid samples were obtained by transabdominal amniocentesis or by amniotomy from 198 pregnant women from weeks 15 to 42. Gestational age was determined by the usual obstetric criteria and by sonography. The amnioceteses performed early in gestation were conducted for the purpose of genetic studies. In late gestation the amnioceteses were carried out to examine the fetal lung maturity when the termination of pregnancy was indicated. Different amniotic fluid samples were obtained at the time of cesarean section or delivery. All samples were immediately centrifuged at 1,200 g for 15 min. The supernatant was heated at 56°C for 10 min.

Russel's Viper Venom

It was supplied as a freeze-dried preparation (Wellcome Reagents Ltd., Beckenham, England) and reconstituted in isotonic saline containing 0.2% ovolecithin.

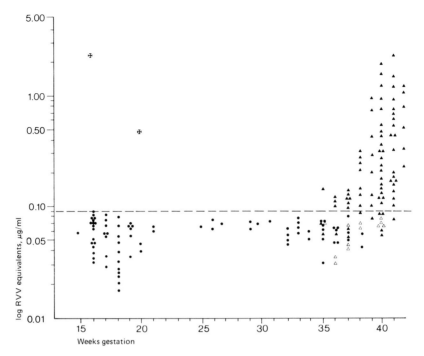

Fig. 1. RVV equivalents of AF in correlation to advancing gestational age. ● = AF obtained by amniocentesis; no delivery within 24 h; ▲, △ = AF obtained when delivery was within 24 h – no RDS (▲): RDS (△);†=AF from dead fetus.

Assay

The mixture of 0.1 ml of AF prepared as described above, 0.1 ml Michaelis buffer (pH 7.34) and 0.1 ml substrate plasma (Merz u. Dade GmbH, München, FRG) was incubated at 37°C for 1 min. The coagulation time was measured with a semiautomatic coagulometer after adding 0.1 ml CaCl$_2$ (0.025 M). Since various charges of substrate plasma had to be used, the measured coagulation times were related to an equivalent of RVV. The equivalents of RVV were quoted from a RVV calibration curve.

Dynamic Surface Properties

These were measured in a Wilhelmy balance [9]. We did not centrifuge these AF samples. The size of the Teflon trough was 13.3 × 1.8 × 1.7 mm. The duration of a cycle was 4 min. Surface tension diagrams were recorded on an x/y plotter. The following criteria were estimated: surface tension at 100% surface and surface tension at 20% of the initial surface and the hysteresis area.

L/S Ratio

Sphingomyelin and lecithin were estimated according to *Hobbins et al.* [7].

Table I. Frequency of RDS after the assessment of lung maturity by estimation of RVV equivalents

RVV equivalents, (µg/ml)	n (total)	n (mature)	n (immature)
≥ 0.09	72	72	0
			(0% false-negative)
< 0.09	23	11	12
		(47,8% false-positive)	(52,2% positive)

Newborns and RDS

In 95 cases AF was obtained at least 24 h before delivery. The results of the AF investigation were compared with the fetal outcome. Additionally, the dynamic surface properties of AF and the L/S ratio were measured in 19 cases. RDS was defined as follows: (1) grunting; (2) sternal and intercostal retractions; (3) reticular granularity in X-ray; (4) tachypnea, and (5) cyanosis in room air.

A diagnosis was not made unless at least criteria 1–3 was observed.

Results

Figure 1 shows the RVV equivalents of AF in correlation to the advancing gestational age. A value of 0.09 µg/ml RVV equivalent was not attained before week 36, whereas a steep increase was seen after week 36 up to values of 2–3 µg/ml. There are 2 values from weeks 16 and 21 exceeding 0.09 µg/ml; these AF samples were obtained from fetuses which were already dead. As illustrated in table I there were 95 newborns of which the pulmonary condition was examined after preceding determination of RVV values in AF. The RVV equivalents were 0.09 µg/ml or more in 72 cases. None of these showed RDS. 23 AF samples were below 0.09 µg/ml; 12 of these babies developped RDS ($\hat{=}$52.2%). The remaining 11 cases showed no RDS, thus revealing false-positive (immature) results ($\hat{=}$47.8%). No false-negative (mature) result was seen. 19 AF samples were examined by estimation of RVV equivalents, by L/S ratio and by dynamic surface tension (Wilhelmy balance). As demonstrated in table II, the different methods for predicting the pulmonary maturity coincided with the clinical outcome in 16 cases. The applied test methods showed contrary results 3 times: No. 7 – the fetal outcome was not documented; No. 13 – the RVV test predicted a transitory RDS, and No. 14 – the RVV value failed in a diabetic mother.

Table II. Assessment of fetal lung maturity by the L/S ratio, surface tension and RVV equivalents

Case No.	L/S ratio	Surface tension	RVV equivalent	Neonatus
1	+	+	+	m
2	–	–	–	
3	–	–	–	
4	–	–	–	
5	+	+	+	m
6	–	–	–	
7	+	+	–!	no report
8		+	+	m
9	–	–	–	
10	+	+	+	m
11	–	–	–	
12	–	–	–	
13	+	+	–!	trans. RDS
14	+	+	–!	m (diabetes)
15	–	–	–	
16	+	+	+	m
17	–	–	–	
18	–	–	–	
19	–	–	–	

+ = mature, – = immature result; m = mature newborn

Discussion

Amniotic fluid accelerates the clotting process due to thromboplastic activity promoting the conversion of prothrombin to thrombin [4]. It is a characteristic of AF to shorten the clotting times of plasma deficient in coagulation factors VIII, IX, XI, XII and VII; in the absence of F X and V, AF does not work in that way [11]. Thus, it can be concluded that AF activates F X directly, since a plasma defective in F V cannot be corrected by F X activation (fig. 2). This procoagulant activity of AF acts in a manner somewhat similar to RVV [10, 12]. In one paper [6] it is reported that the procoagulant property of AF increases with gestational age as measured by whole blood clotting time. The need of using different blood samples impacts the comparability of this method. Thus, it was not possible to get any information about correlations between the coagulation assay and bio-

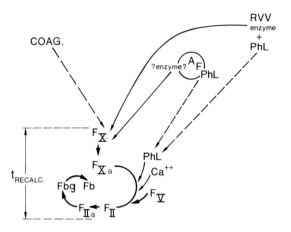

Fig. 2. Interference of amniotic fluid with coagulation mechanisms. PhL = Phospholipid.

chemical or clinical parameters for fetal lung maturity. Because of the similar procoagulant functions in AF and RVV, the procoagulant activity of AF has been quantitated in terms of RVV equivalents in the assay described above. In our assay the AF was heated to inactivate triggered clotting factors present due to blood contamination. The addition of a buffer causes a constant pH and permitted reproducible results. The concentration of Ca^{2+} ions of AF did not play any role in the recalcification time. As demonstrated in figure 1 the procoagulant activity in AF shows no increase from weeks 15 to 35, whereas a clear rise was noted from week 36. Two AF samples obtained from dead fetuses in week 16 and 21 exhibit values usually found at term. This can be explained by an accumulation of thromboplastic material derived from necrotic fetal tissue [2].

Analyzing these results for clinical significance a RVV equivalent of 0.09 µg/ml was found empirically to be a limiting value for fetal pulmonary maturity (table I). When this RVV value was reached or even passed, no RDS was observed in the newborn period, the predictive value of a mature (negative) result was 100%. A RVV value below 0.09 µg/ml revealed false immature (positive) results in 47.8%, exhibiting an efficiency comparable to the L/S ratio [8]. In the use of the definition, a positive test is one that is indicative of disease (in our discussion fetal lung immaturity or RDS) and a negative test is one which is indicative of health (fetal lung maturity). This use of the terms conforms to the common usage in other areas of medicine and is, we think, logical, but it is directly opposite to the usage of the terms positive and negative in most of the obstetric literature dealing with tests for fetal lung maturity.

As illustrated in table II the results of the 3 laboratory tests – (1) L/S ratio; (2) surface tension; (3) coagulation assay – were compared versus the neonatal outcome. Whereas all three tests succeeded in predicting the pulmonary condition in 16 cases, there were 3 cases in whom the results of the different methods did not coincide. Once, the information about the neonatus was not available. In 2 other cases the result of the RVV assay was not in agreement with the L/S ratio and surface tension. In 1 of these 2 cases, a transitory RDS occured as predicted by the RVV test. The other baby was mature despite a low RVV value, but the mother was diabetic.

It has to be emphasized that the finding of an immature test for fetal lung maturity cannot be proved by the neonatal outcome in the most cases, because each obstetrician would try to manage this situation in order to deliver a mature baby. Because of this fact the study presented here cannot demonstrate a sufficient number of RDS for evaluating the exact efficiency of the coagulation test. But it is of great interest that a mature test indicates the pulmonary maturity with a predictive value of 100%. It can be stated that the use of more than one test enhances the degree of confidence for predicting the lung condition and further investigation should elucidate the role of the coagulation factors in the amniotic fluid.

Summary

With advancing gestational age the procoagulant activity increases in the amniotic fluid. This procoagulant activity was measured by use of a modified recalcification time under standardized conditions. The results were expressed as equivalents of RVV which acts in the same way as amniotic fluid. After gestational week 35, RVV values beyond 0.09 µg/ml indicate the fetal lung maturity. Higher than that value no RDS was observed.

References

1   Avery, M.E. and Mead, J.: Surface properties in relation to atelectasis and hyaline membrane disease. Am. J. Dis. Child. 97: 517–522 (1959).

2   Beller, F.K. and Uszynsky, M.: Disseminated intravascular coagulation in pregnancy. Clin. Obstet. Gynec. 17: 250–278 (1974).

3   Clements, J.A.: Surface tension of lung extracts. Proc. Soc. exp. Biol. Med. 95: 170–174 (1957).

4   Courtney, L.D. and Allington, M.: Effect of amniotic fluid on blood coagulation. Br. J. Haemat. *22:* 353–355 (1972).

5   Gluck, L.; Motoyama, E.K.; Smiths, H.L., and Kulovich, M.W.: The biochemical development of surface activity in mammalian lungs. I. The surface active phospholipids, the separation and distribution of surface active lecithin in the lung of the developing fetus. Pediat. Res. *1:* 237–246 (1967).

6   Hastwell, G.B.: Accelerated clotting time. An amniotic fluid thromboplastic activity index of fetal maturity. Am. J. Obstet. Gynec. *131:* 650–656 (1978).

7   Hobbins, I.C.; Brock, W.; Speroff, L.; Anderson, G.G., and Caldwell, B.: L/S ratio in predicting pulmonary maturity *in utero.* Obstet. Gynec. *39:* 660–664 (1972).

8   Kubli, F.; Lorenz, U. und Rüttgers, H.: Fruchtwasseruntersuchungen zur intrauterinen Reifegradbestimmung. Gynäk. Rdsch. *14:* suppl. 1, pp. 10–20 (1974).

9   Müller-Tyl, E.; Lempert, J.; Steinbereithner, K., and Benzer, H.: Surface properties of the amniotic fluid in normal pregnancy. Am. J. Obstet. Gynec. *122:* 295–300 (1975).

10   Phillips, L.L. and Davidson, E.C.: Procoagulant properties of amniotic fluid. Am. J. Obstet. Gynec. *113:* 911–919 (1972).

11   Stolte, L.S.; Lim, H.T.; Van Arkel, C.; Eskes, T.K.A.B., and Van Kessel, H.: Activation of coagulation, stasis and the amniotic fluid embolism syndrome. Eur. J. Obstet. Gynec. *6:* 219–224 (1971).

12   Williams, W.J. and Esnouf, M.P.: The fractination of Russells-viper (Vipera russellii) venom with special reference to the coagulant protein. Biochem. J. *84:* 51–62 (1962).

13   Yaffe, H.; Eldor, A.; Hornshtein, E., and Sadovsky, E.: Thromboplastic activity in amniotic fluid during pregnancy. Obstet. Gynec. *50:* 454–456 (1977).

Dr. med. W. Leucht, Universitäts-Frauenklinik, Vossstrasse 9, D–6900 Heidelberg (FRG)

Prog. Resp. Res., vol. 15, pp. 133–135 (Karger, Basel 1981)

Relationship between Fetal Heart Rate Patterns and Changes in the Lung Phospholipid Profile during Labor

M.J. Whittle, K.S. Koh, E.H. Hon, M. Kulovich and L. Gluck

Division of Maternal-Fetal Medicine, Department of Obstetrics and Gynecology,
University of Southern California School of Medicine, Women's Hospital, Los Angeles
County-University of Southern California Medical Center, Los Angeles, and
Department of Pediatrics, University of California, San Diego, La Jolla, Calif.

The association of severe intrapartum asphyxia and the development of respiratory distress syndrome (RDS) in the neonante have been previously observed [1]. Further, the appearance of abnormal fetal heart rate (FHR) patterns and an increased incidence of RDS in the low birth weight neonate has been described [2]. The reason for these observations is not clear, but we were interested to see if changes in surfactant status were linked with abnormality in FHR during labor.

Material and Methods

38 patients in spontaneous labor were studied. Aliquots of amniotic fluid were drawn through an intrauterine catheter at approximately 2-hour intervals for as long as possible during the labor. The fluids were analyzed using a two-dimensional, thin-layer chromatography technique allowing estimation of the phospholipids present (lecithin, sphingomyelin, phosphatidylinositol, phosphatidylglycerol).

Without prior knowledge of the amniotic fluid results, the FHR patterns were analyzed using a modification of the scoring system used by *Schifrin and Dame* [3].

Results

Three groups of patients were defined; in group I the L/S rose, in group II the L/S remained unchanged, and in group III the L/S fell. The changes occurring in the concentrations of phosphatidylinositol and phosphatidylglycerol did not follow a consistent pattern.

Group I patients were found to have significantly shorter labors than group III (p < 0.01). Although more abnormalities in FHR patterns appeared to occur in group III the differences were not significant from those in group I. However, fetal heart rate baseline variability was significantly depressed in group III (p < 0.02 > 0.01) when compared to group I. This appeared to be independent of the type of analgesia used during labor. When the FHR baseline variability was assessed for the 30 min pior to amniotic fluid sampling, it was found that when variability was depressed the L/S ratio was significantly lower (p < 0.001).

Discussion

The observation of changes in surfactant status in association with FHR abnormalities has not previously been made. Although the development of RDS is more common in the asphyxiated neonate, it is not clear whether our observations are of a pathological or physiological process. Although depression of FHR baseline variability has been associated with fetal hypoxemia [4], in the present study there were not other features suggesting fetal asphyxia.

The possibility remains that we are observing a physiological process, the autonomic nervous system influencing both fetal heart rate variability and the rate of release of surfactant.

Summary

Serial samples of amniotic fluid were taken from 38 patients in spontaneous labor. The phospholipid profile of the amniotic fluid was determined and compared with the characteristics of the fetal heart rate (FHR) patterns. The changes in the lecithin/sphingomyelin (L/S) ratio were significantly correlated with the length of labor and FHR variability. It is proposed that the autonomic nervous system activity may be responsible for both changes in L/S ratio and FHR variability.

References

1   Donald, I.R.; Freeman, R.K.; Goebelsmann, U.; Chan, W.H., and Nakamura, R.M.:
     Clinical experience with amniotic fluid L/S ratio. Am. J.Obstet. Gynec. *115:* 547–552
     (1973)

2   Martin, C.B.; Siassi, B., and Hon, E.H.: Fetal heart rate patterns and neonatal death in low birthweight infants. Obstet. Gynec. *44:* 503–510 (1974).
3   Schifrin, B.S. and Dame, L.: Fetal heart rate patterns. Prediction of Apgar score. J. Am. Med. Ass. *219:* 1332–1325 (1972).
4   Paul, R.H.; Suidan, A.K.; Yeh, S.-Y; Schifrin, B.S., and Hon, E.H.: Clinical fetal monitoring. VII. The evaluation and significance of intrapartum baseline FHR variability. Am. J.Obstet. Gynec. *123:* 206–210 (1975).

M.J. Whittle, MD MRLOG, Department of Midwifery, The Queen Mother's Hospital, Glasgow, Scotland.

Prog. Resp. Res., vol. 15, pp. 136–140 (Karger, Basel 1981)

Surfactant Metabolism in Acute Pancreatitis

R.B. Passi and F. Possmayer

Departments of Surgery, Obstetrics and Gynaecology, Physiology and Biochemistry, University Hospital, University of Western Ontario, London, Ont.

Introduction

Respiratory complications are the most common seen in all types of pancreatitis and 20 of 24 patients with acute haemorrhagic pancreatitis had some form of pulmonary complications. Of the pulmonary complications atelectasis and pulmonary effusion are seen in more than 50% of cases. However, the adult respiratory distress syndrome developed in 25% of the cases and this complication had a uniform mortality rate of 100% and became the major therapeutic problem [5].

The development of the adult respiratory distress syndrome is seen clinically when the pO_2 decreases, the respiratory rate increases and the chest x-ray shows bilateral pulmonary infiltration. The aetiology of this pulmonary distress syndrome seen in acute pancreatitis has not been clearly established. When one considers the complexity of the fluid, electrolyte, metabolic and enzymatic problems in acute pancreatitis, a single aetiology can hardly be expected to account for this rather complex problem. Simple fluid overload, heart failure due to a myocardial depressant factor, pulmonary hypertension due to disseminated intravascular coagulation, vasoactive substances and endotoxins, may play a part in some of these patients but are likely not the main factors in view of normal central venous, pulmonary artery and pulmonary capillary wedge pressures and normal pulmonary vascular resistance in most patients with acute adult respiratory distress syndrome in acute pancreatitis [8]. Hyperlipidaemia with the formation of free fatty acids in the pulmonary alveoli may play a significant role [4].

The enzymatic destruction of surfactant and subsequent alveolar collapse is an attractive hypothesis as one main aetiologic factor in the development of acute respiratory distress syndrome in pancreatitis. Phospholipase A_2 has been demonstrated to increase ten-fold in patients with acute pancreatitis [9]. The administration of exogenous phospholipase A has been shown to have an affinity to the alveolar membrane [3]. Surfactant is rich in phospholipids particuly lecithin that may be broken down to phospholipase A_2 to form lysolecithin which may act as an anti-surfactant or cellular toxin and be responsible for the development of interstitial alveolar oedema and subsequent pulmonary distress. This hypothesis was studied in the experimental laboratory.

Material and Methods

Adult mongrel dogs weighing 20–25 kg were anaesthetized with sodium pentothal induction, maintained with 1% halothane and room air and ventilated endotracheally with a tidal volume of 15 cm^3/kg body weight. Intravenous normal saline was given at 10 cm^3/kg/h for fluid maintenance.

Acute haemorrhagic pancreatitis was produced in the dogs by transduodenal low-pressure injection of 0.5 ml/kg of autologous bile obtained from the gallbladder into the main pancreatic duct. The common duct and accessory pancreatic duct was ligated. The development of pancreatitis was verified by first noting bile staining of the pancreas at the time of injection and finally by pathological evidence of acute haemorrhagic pancreatitis at the termination of the experiment after 12 h. Control dogs were similarly anaesthetized, ventilated and subjected to laparotomy at which time bile was aspirated from the gallbladder, a duodenotomy was done and the pancreatic duct was isolated but not injected. In addition, the accessory pancreatic duct and common duct was not ligated. The control animals were also ventilated for 12h and received the same volume of intravenous saline. Pulmonary function studies were determined at 4-hour intervals which included tidal volume, minute volume and functional residual capacity. In addition, the arterial partial pressures of oxygen and carbon dioxide were determined as well as the arterial pH, bicarbonate and base excess. Serum was also obtained at 4-hour intervals for the determination of phospholipase A activity according to the method of Zieve and Vogel [9].

At 12 h, the dogs were sacrificed and the pulmonary alveoli were immediately lavaged with three successive rinses of 500 ml of normal saline. The collected suspensions were centrifuged and the lipid extracted with chloroform: methanol. The phospholipid composition was then assayed by thin-layer chromatography either by the method of Rouser et al. [6] or Broekhuyse [2]. Phospholipase A activity was assessed by the method of Zieve and Vogel [9]. The surface activity of the lung wash phospholipid was assessed on an artificial alveolus machine as described by Adams and Enhorning [1].

Results

The results ensuing are at variance with each other. Control animals demonstrated an improvement in respiratory function during the 12-hour period as the PO_2 rose from 86 to 102 mm Hg. An increase in metabolic acidosis occurred as bicarbonate fell from 18 to 15 and base excess from –5 to –9. However, the pH remained stable at 7.34 as the pCO_2 fell from 35 to 29 mm Hg. In contrast, dogs with pancreatitis had an initial improvement in pulmonary function followed by deterioration at the end of the experiment. During the first 8 h the pO_2 rose from 88 to 101 mm Hg. but by 12 h had fallen to 87 mm Hg. There was a greater increase in metabolic acidosis as evidenced by a fall in bicarbonate from 19 to 14 and base excess from –6 to –12. The pH changed slightly remaining around 7.33 while the pCO_2 dropped from 36 to 27 mm Hg. The functional residual capacity rose by 12% in control animals during the experiment and in those dogs subjected to pancreatitis rose by 33% from 4 to 8 h during the experiment but by 12 h had returned to pre-operative values.

At 12 h as the arterial partial pressure of oxygen and the functional residual capacity of lung were decreasing, it may be that pulmonary insufficiency was just beginning and the experiment was arbitrarily terminated a few hours too early.

Phospholipid yield of lung wash at 12 h was greater in 5 control animals as compared to 5 animals with acute pancreatitis. The phospholipid composition was similar though there was a drop in phosphatidylcholine and increase in lysophosphatidylcholine in those animals with acute pancreatitis. When the serum and lung wash of these 5 animals were incubated under conditions of phospholipase A activity there was a marked decrease in the lecithin and increase in lysolecithin. Unfortunately, however, there was no significant difference of the lung wash materials to lower surface tension in either control or experimental groups using the pulsating bubble machine for surface activity.

A second group of animals involved 4 control, 7 experimental animals and 10 dogs not subjected to any experimentation. The control and experimental animals were treated as earlier described for a 12-hour period. Gross changes are seen to occur in three areas. The yield of phospholipid was significantly reduced in the experimental group. Of more significance is the increase in the lysolecithin fraction in the experimental versus the controlled group. The control group's values are very much in keeping with the values of normal dogs. In addition in these 7 dogs there was a clear increase in the surface activity of the lung wash in experimental animals.

Comments

The data indicate that in acute experimental pancreatitis, recovery of surfactant is reduced while the lysolecithin is increased. It correlated well with a decreased ability of the lung wash from experimental groups to lower surface tension of a pulsating bubble, indicating poor quality surfactant. Though the surface activity changes in control and experimental groups have not been altered in two different experimental environments, we have demonstrated deteriorating pulmonary function in acute experimental pancreatitis as well as a decrease in phospholipid yield from lung wash, increase in phospholipase A activity in both serum and lung wash and impaired surface tension reducing capabilities of lung wash in one experimental group. We thus conclude that in acute pancreatitis lysolecithin produces an antisurfactant activity either by injury to the surfactant itself or the alveolar type 2 cell to produce surfactant. This results in an increasing pulmonary surface tension. Hydrolysis of lipids in the capillary bed of the lungs by pancreatic phospholipase A2 is strongly suggested but not proven to date, and increased lysolecithin in surfactant results in atelectasis. The comined effects would lead to pulmonary oedema and the development of the adult respiratory distress syndrome in acute pancreatitis.

Summary

In one experimental group of acute pancreatitis, pulmonary surfactant has been reduced and lysolecithin increased. In a second group, phospholipase A_2 activity is present but did not alter pulmonary surfactant. Pulmonary function deteriorated in both groups. We concluded, however, that in acute pancreatitis, lysolecithin produces an antisurfactant activity by injury to the surfactant or the alveolar type 2 cell. Pancreatic phospholipase A_2 is suggested but not proven to cause the degradation of surfactant and hydrolysis of pulmonary lipids.

References

1   Adams, F.H. and Enhorning, G.: Surface properties of lung extracts. 1. A dynamic alveolar model. Acta physiol. scand. *68:* 23 (1966).

2   Broekhuyse, R.M.: Quantitative two-dimensional thin-layer chromatography of blood phospholipids. Clinica chim. Acta *23:* 457 (1969).
3   Gennaro, J.: Studies on the distribution of radio-iodine labelled snake venom. Int. Symp. on Animal Toxins, Atlantic City 1966.
4   Kimura, T.; Toung, J.K.; Margolis, S.; Permutt, S., and Camperon, J.L.: Respiratory failure in acute pancreatitis: a possible role for triglycerides. Ann. Surg. *189:* 509 (1979).
5   Passi, R.B.: Unpublished data.
6   Rouser, G.; Siakotos, A.N., and Fleischer, S.: Quantitative analysis of phospholipids by thin layer chromotography and phosphorous analysis of spots. Lipids *1:* 85 (1966).
7   Tierney, D.F. and Johnson, R.P.: Altered surface tension of lung extracts and lung mechanics. J. appl. Physiol. *20:* 1253 (1965).
8   Warshaw, A.L.; Lesser, P.B.; Rie, M., and Cullen, D.J.: The pathogenesis of pulmonary edema in acute pancreatitis. Ann. Surg. *182:* 505 (1975).
9   Zieve, L. and Vogel, W.C.: Measurement of lecithinase A_2 in serum and other body fluids. J.Lab. clin. Med. *52:* 586 (1961).

R.B. Passi, MD, Departments of Surgery, Obstetrics and Gynaecology, Physiology, and Biochemistry, University Hospital, University of Western Ontario, London, Ontario N6A 5A5 (Canada)

Discussion

In his answer to a question by *von Wichert* as to whether the phospholipase A2 shown in lung washes comes from the pancreas or from the lung itself, *Passi* stated that that was not clear until now. He had tried to isolate that enzyme. *Morley* mentioned that the lung surfactant from pigs used for pancreas transplantation experiments was not as effective as that from normal pigs. *Rüfer,* referring to experiments on shock lung, induced by thrombin infusion, stated that in shocked lungs there was always a large amount of lysolecithin that was not present in normal lungs.

Prog. Resp. Res., vol. 15, pp. 141–147 (Karger, Basel 1981)

Lung Phospholipid Metabolism after Smoke Exposure in Rabbits[1]

W. Meyer, A. Burkhardt, B. Klenke, H. Vogts, A. Wilke and P. von Wichert

I. Medizinische Klinik des Universitätskrankenhauses, and Institut für Pathologie des Universitätskrankenhauses Hamburg-Eppendorf, Hamburg.

Introduction

The retention of the normal structure and thus of the normal function of the alveoli depends, at least in part, on an intact surface-tension-reducing surfactant [9,12,13]. Therefore, the occurrence of some chronic lung deseases like, for instance, emphysema may be due to alterations in surfactant activity. On the other hand, it is well known that chronic smokers suffer more frequently from emphysema than nonsmokers. Because of this relationship the question arises as to whether inhalation of cigarette smoke might lead to an altered synthesis of surfactant. Although several reports [5,7,14,19] have shown that cigarette smoke decreased surfactant activity in bronchial washings, there is only contradictory information about its influence upon surfactant metabolism measured by incorporation of radioisotopically labeled precursors into canine lung phospholipids *in vivo* [2,10].

In order to find an explanation for the influence of cigarette smoke on the biosynthesis of phosphatidylcholine in rabbit lung we investigated two distinct alterations possibly depending on the duration of smoke exposure: changes in the ability of lung slices to incorporate fatty acids and choline into phosphatidylcholine and changes in the amount and in the composition of phosphatidylcholine in lung tissue.

[1] Supported by the 'Forschungsrat Rauchen und Gesundheit'

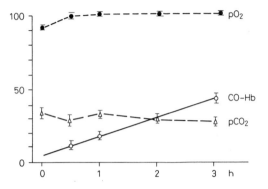

Fig. 1. Assessment of blood gases (pO_2, pCO_2 in mm Hg) and CO-Hb (%) during smoke exposure for 3 h.

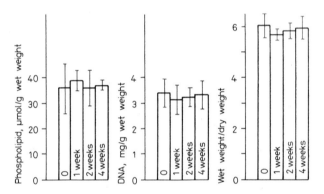

Fig. 2. Phospholipid content, DNA content, and wet weight/dry weight ratios of rabbit lung after 1, 2, and 4 weeks of smoke exposure.

Materials and Methods

Rabbits of both sexes weighing between 2.8 and 3.5 kg were used. They had free access to food and water. Test cigarettes contained 14 mg condensate and 0.8 mg nicotine. Smoke exposure was achieved by a smoking machine type 'Hamburg 2' according to *Dontenwill* [6]; the animals were kept under standard conditions in an atmosphere of six-fold diluted cigarette smoke. CO-Hb, pO_2 and pCO_2 were assessed over a period of 3 h (fig. 1). At the end of this period, the rabbits were seriously ill. So, in order to keep CO-Hb levels in the test animals below 20%, the experiments were carried out by two exposures of 0.5 h per day. After 1, 2 and 4 weeks, the animals were sacrificed and their lungs removed and investigated. Estimation of DNA content, lipid extraction and analysis of phospholipid content were carried out by suitable methods [3, 4, 8], individual phospholipids were separated by thin-layer chromatog-

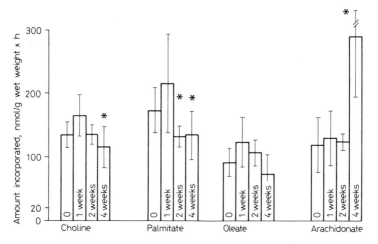

Fig. 3. *In vitro* incorporation rates of different precursors into rabbit lung phosphatidyl-choline after 1, 2, and 4 weeks of smoke exposure.

raphy on HPTLC plates in $CHCl_3/CH_3OH/CH_3COOH/H_2O$ 60:30:8:4 (vol/vol), visualized by iodine vapor and quantified by phosphate estimation [16]. Determination of the fatty acid composition of phosphatidylcholine was achieved by gas chromatography (GC) after isolation by preparative TLC under nitrogen and preparation of fatty acid methylesters with BF_3/CH_3OH [15]. GC analysis was performed on a Hewlett-Packard model 5700 A gas chromatograph equipped with a flame ionization detector, an integrator, and a 6-ft glass-column packed with 2.5% DEGS on Chromosorb G AW/DMCS (80–100 mesh). Incorporation rates of ^{14}C-labeled palmitate, oleate and arachidonate (bound to BSA, specifec radioactivity: 1.03μCi/μmol) and of (methyl-^{14}C) choline (specific radioactivity: 0.22 μCi/μmol) were measured by incubation of lung slices and radioactive precursors (1 μmol/ml) in a total volume of 2.5 ml Krebs-Ringer solution for 1 h at 37°C [11]. Incubation was stopped by homogenization in $CHCl_3/CH_3OH$ 2:1 (vol/vol) [8] and phospholipids were separated and analyzed as described earlier [20].

Results and Discussion

As shown in figure 2, cigarette smoke does not lead to any significant alteration in phospholipid and DNA content or in wet weight/dry weight ratio in rabbit lung over the observed periods up to 4 weeks. Histological investigations exhibited only small dystelectatic regions and some incorporation of pigments. On the other hand, the ability of lung tissue to incorporate precursors into phosphatidylcholine is affected by cigarette smoke (fig. 3). *In vitro* incorporation rates for all precursors are increased after

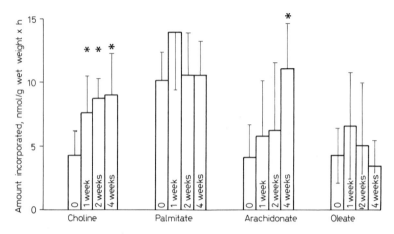

Fig. 4. In vitro incorporation rates of different precursors into rabbit lung sphingomyelin after 1, 2, and 4 weeks of smoke exposure.

1 week of smoke exposure. This effect could also be observed in the 3-hour experiments, but these results are not included because those rabbits could not be regarded as physiological. Within the next weeks of further smoke exposure this trend changes and leads to depressed oleate incorporation and to significantly depressed choline and palmitate incorporation. The only exception is a more than two-fold increase in arachidonate incorporation. Figure 4 shows the *in vitro* incorporation rates of the same precursors into the other choline-containing phospholipid, sphingomyelin. In the case of fatty acids the pattern is similar to that of phosphatidylcholine, choline incorporation, however, appears to be significantly increased.

The results of phospholipid analysis are shown in figure 5. There is a constant decrease in percent phosphatidylcholine of total phospholipids during the experimental period. This trend is rather obvious and is reflected by a corresponding increase in the other non-choline-containing phospholipids. Lysophosphatidylcholine and sphingomyelin do not seem to be affected. The fatty acid pattern of lung phosphatidylcholine shown in figure 6 displays a significant decrease in the palmitate content after 2 and 4 weeks of exposure to smoke. This decrease is compensated by a corresponding increase in total unsaturated fatty acids, especially of palmitoleate and linoleate. Palmitate, however, is required for remodeling steps of *de novo* synthezised phosphatidylcholine to form the disaturated species

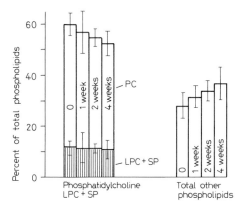

Fig. 5. Phospholipid composition in rabbit lung after 1, 2, and 4 weeks of smoke exposure. LPC = Lysophosphatidylcholine; PC = phosphatidylcholine; SP = sphingomyelin.

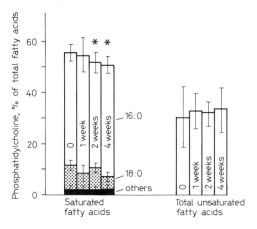

Fig. 6 Fatty acid pattern of rabbit lung phosphatidylcholine after 1, 2, and 4 weeks of smoke exposure.

[1, 17]. Other phosphatidylcholines than 1,2-dipalmitoyl-sn-glycero-3-phosphocholine have been shown to be less effective in their surface-active properties [18].

These results seem to indicate that the decrease of surfactant activity often observed in the lungs of chronic smokers is, at least in part, due to an impaired phosphatidylcholine biosynthesis. This could be demonstrated by a depressed choline incorporation into lung phosphatidylcholine *in vitro* and by a lowered phosphatidylcholine content of total phospholipids in the lungs of rabbits exposed to cigarette smoke. Because of the fact that sphingomyelin biosynthesis is not reduced in the animals, cholinephosphotransferase might be the affected enzyme that is responsible for the impaired phosphatidylcholine biosynthesis. The depressed palmitate incorporation into phosphatidylcholine after 4 weeks of smoke exposure, which is in agreement with other experiments [2], is mirrored by the fatty acid pattern of phosphatidylcholine of rabbit lung after the same period of smoke exposure. The significantly decreased percentage of palmitate indicates a lowered content of disaturated phosphatidylcholine, which results in a deteriorated surfactant activity [18].

So we can summarize two effects of cigarette smoke in rabbit lung: a reduction in phosphatidylcholine combined with a degradation of its quality as a surface-active material because of its lowered palmitate content.

References

1   Akino, T.; Abe, M., and Arai, T.: Studies on the biosynthetic pathways of molecular species of lecithin by lung slices. Biochim. biophys. Acta *248:* 274–281 (1972).

2   Balint, J.A.; Bondurant, S., and Kyriakides, E.: Lecithin biosynthesis in cigarette smoking dogs. Archs. int. Med. *127:* 740–747 (1971).

3   Bartlett, G.R.: Phosphorus assay in column chromatography. J. biol. Chem. *234:* 466–468 (1959).

4   Burton, K.: Determination of DNA content. Biochem. J. *62:* 315–321 (1962).

5   Cook, W.A. and Webb, W.R.: Surfactant in chronic smokers. Ann. thor. Surg. *2:* 327–333 (1966).

6   Dontenwill, W.: Experimental investigations on the effect of cigarette smoke inhalation on small laboratory animals; in: Hanna, Nettesheim and Gilberts, Inhalation carcinogenesis. AEC Symp. Ser. vol. 18, pp. 389–412 (1970).

7   Finley, T.N. and Ladman, A.J.: Low yield of pulmonary surfactant in cigarette smokers. New Engl. J. Med. *286:* 223–227 (1972).

8   Folch, J.M.; Lees, M., and Sloane Stanley, G.H.: A simple method for the isolation and purification of total lipids from animal tissue. J. biol. Chem. *226:* 497–509 (1957).

9   Frosolono, M.F.: Lung; in: Snyder, Lipid metabolism in mammals II, pp. 1–38 (Plenum Press, New York 1977).

10   Giammona, S.T.; Tocci, P., and Webb, W.R.: Effects of cigarette smoke on incorporation of radioisotopically labeled palmitic acid into pulmonary surfactant and on surface activity of canine lung extracts. Am. Rev. resp. Dis. *104:* 358–367 (1971).

11   Gilder, H. and McSherry, C.K.: An improved method for measuring incorporation of palmitic acid into lung lecithin. Am. Rev. resp. Dis. *106:* 556–562 (1972).

12   Goerke, J.: Lung surfactant. Biochim. biophys. Acta *344:* 241–261 (1974).

13   King, R.J.: The surfactant system of the lung. Fed. Proc. *33:* 2238–2247 (1974).

14   Miller, D. and Bondurant, S.: Effect of cigarette smoke on the surface characteristics of lung extracts. Am. Rev. resp. Dis. *85:* 692–696 (1962).

15   Morrison, W.R. and Smith, L.M.: Preparation of fatty acid methyl esters and dimethy-lacetals from lipids with boron fluoride-methanol. J. Lipid Res. *5:* 600–608 (1964).

16   Veerkamp, J.H. and Broekhuyse, R.M.: Techniques for the analysis of membrane lipids; in: Maddy, Biochemical analysis of membranes, pp. 252–282 (Chapman & Hall, London 1976).

17   Vereyken, J.M.; Montfoort, A., and van Golde, L.M.G.: Some studies on the biosynthesis of the molecular species of phosphatidylcholine from rat lung and phos-phatidylcholine and phosphatidylethanolamine from rat liver. Biochim. biophys. Acta *260:* 70–81 (1972).

18   Watkins, J.C.: The surface properties of pure phospholipids in relation to those of lung extracts. Biochim. biophys. Acta *152:* 293–306 (1968).

19   Webb, W.R.; Cook, W.A.; Lanius, J.W., and Shaw, R.R.: Cigarette smoke and surfactant. Am. Rev. resp. Dis. *95:* 244–247 (1967).

20   Wichert, P. von; Wilke, A. und Gärtner, U.: Einbau von Palmitat-1-^{14}C in Lecithin und Phospholipidgehalt in normalen und mikroembolisierten Kaninchenlungen. Modell-studie zur sogenannten Schocklunge. Anaesthesist *24:* 78–83 (1975).

W. Meyer, MD, I. Medizinische Klinik des Universitätskrankenhauses, D-2000 Hamburg-Eppendorf (FRG)

Discussion

Robertson asked whether macrophages, known to be increased after smoke exposure, could interfere with the results. *Meyer* answered that there was only a slight increase in the alveolar macrophages after 4 weeks. *Reifenrath* drew attention to the fact that surfactant may be much more important at the bronchial level than at the alveolar level. At the bronchiolar level, expecially in smoke exposure, deficiency of the surfactant leads to earlier closure of the small bronchi because of increased adhesion of bronchoilar walls which is normally prevented by the surfactant. *Tierney* suggested that the increase in arachidonic incorporation could allow lipid peroxidation which could be damaging to lung tissue. These peroxidation products were reported to inhibit α_1-antitrypsin, thus lung damage after smoke exposure may not be directly related to the surfactant. *Reiss* confirmed that activated alveolar macrophages incorporate arachidonic acid into PC and PE at a very rapid rate; it is subsequently hydrolyzed by phospholipase A to prostaglandines which may cause the injury. Arachidonic acid is also a precursor of a slow reacting substance. *Clements* remarked that other mechanisms besides surfactant could be involved and suggests that it has effects on bacterial killing by macrophages and on the immunoligical activity of leukocytes.

Prog. Resp. Res., vol. 15, pp. 148–154 (Karger, Basel 1981)

Clinical Importance of Surfactant Defects in Perinatology

J.W. Dudenhausen

Department of Obstetrics, Berlin-Neukölln, Unit of Perinatal Medicine, Free University of Berlin, Berlin

The membrane syndrome causes the most severe disturbance of the function of the lungs in the newborn, in particular in the premature newborn. It is the most frequent cause of death of immature infants with a frequency of about 10%. The shorter the gestation period, the more frequently the syndrome occurs. In spite of very modern intensive neonatal supervision, the mortality rate of infants with the membrane syndrome is still 20%.

The membrane syndrome is a variably strong pulmonary insufficiency occurring during the first hours of life, generally starting after a free interval, exhibiting as its main clinical symptoms tachypnea and dyspnea with inspiratory retraction and expiratory groaning. Thorax roentgenograms show diffuse reticular granular solidification and also an air bronchogram.

The name 'membrane syndrome' comes from hyaline membranes, that is flabby eosinophilic membranes, which coat the patent alveoli. More important functionally than these membranes, which play a purely subsidiary role, is the disturbance in the development of the majority of the alveoli. These are deflated, that is not ventilated, but still perfused. An intrapulmonary right to left shunt occurs.

A membrane syndrome can have various etiologies. In the premature newborn, pulmonary immaturity is the most important etiological factor.

Intrauterine hypoxia and acidosis can occur as further etiological factors in mature and immature newborns alike.

Morphological and Biochemical Maturation

The development of the lungs begins with the evagination of the celenteron, when the embryo is 26 days old, through the glandular stage, at about the 16th week, with plump bronchial tree branches and also a wide, loose mesenchymal coat and the canalicular type of lung, with division and differentitation of the bronchioles; from the 16th week up to about the 24th week, the glandular type of lung is formed through further racemose proliferations.

The acini can be regarded as the immediate forerunners of the mature pulmonary alveoli. Passage to the alveolar lung type starts from about the 25th to 27th week. This period is characterized by the displacement of the placement tissue, the expansion of the pulmonary capillaries approaching the alveolar lumen and the maturation of the type II and I alveolar cells. Their differentiation can already start from about the 22nd week. Next, type II alveolar cells receive globular inclusions which become increasingly lamellar as a sign of maturity. Biosynthesis of the lung surfactant takes place in the secretory device of the type II alveolar cells. The lamellar inclusion bodies serve as an intracellular surfactant reservoir; they empty their contents into the alveoli and in this way form the surfactant layer. The surfactant is removed from the air passages by vibrating epithelium and must therefore be continuously replaced.

In the majority of cases, from about the 26th week of gestation according to *Vogel* [14], not only tissular but also cellular prerequisites must be present for the onset of exogenic lung maturation.

Biosynthesis of the lecithins takes place almost exclusively in two enzymatic processes, whose importance typically changes during the course of pregnancy:

Early process path: in an immature fetus (22nd to 35th week), the production of palmityl-myristyl-lecithin dominates. This synthesis is compromised much more strongly by acidosis, hypoxia and hypercapnia than the dipalmityl-lecithin synthesis described below.

Later process path: from about the 35th week onwards, the production of dipalmityl-lecithin is strongly activated. This can be recognized by the steep increase of lecithin in the amniotic fluid at this time.

Diagnosis of Lung Maturity

In order to recognize a membrane syndrome prenatally and to prevent it, lung maturity can be examined by determining the surfactant concentration, and/or part of the surfactant, i.e. the lecithin, in the amniotic fluid. Lecithin synthesis begins at about the 18th to 24th week of gestation and increases during the course of pregnancy.

Amniotic fluid is obtained by transabdominal amniocentesis; the operation is technically almost without difficulties.

Various physical and biochemical methods serve to assess the lipid concentration in amniotic fluid samples, but none has yet emerged as the ultimate standard. Most importance is awarded to the determination of the lecithin/sphingomyelin ratio according to Gluck, to the direct determination of lecithin and also to the determination of surface characteristics by means of Clements's foam test. We use a combination of both methods, mainly because the accuracy of the foam test is limited [9].

Antenatal Promotion of Lung Maturation

According to the literature, an acceleration of lung maturation is found in cases of gestosis, hypertensive illnesses or previous premature rupture of the membranes. In these cases, the lung maturity test is frequently positive before the 34th week of gestation. On the other hand, in cases where the mother is diabetic, one can reckon with a delay in lung maturation. Sometimes, infants born shortly before term, and apparently not at risk, have a membrane syndrome. In such cases hereditary factors can be responsible [7].

Reliable clues point at a glucocorticoid-regulated fetal lung maturation. At 35 weeks, the corticoid level in the amniotic fluid and in the umbilical cord blood doubles or triples [5]. At the same time, the biochemical maturation of the 'surfactant' occurs. Furthermore, premature newborns with the membrane syndrome have a lower cortisol level in umbilical cord blood than the full-term control group [11]. Corticoid regulation is connected to a cytoplasmic carrier protein and to successive activation of specific genes in the alveolar epithelial cells. Glucocorticoid increases the enzyme activity of the choline-phosphorus-transferase in the fetal lung and stimulates the accumulation of surfactant complexes and their liberation [4].

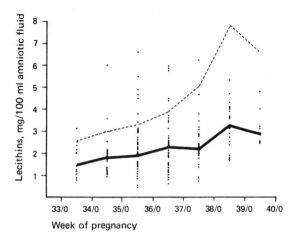

Fig. 1. Medians of the lecithin contents of the amniotic fluid in the group on long-term β-mimetic therapy (——— ; 157 determinations in 60 pregnant women) and in the control group (– – –; 211 determinations in 190 pregnant women) according to the week of pregnancy.

The present results of lung maturation achieved by drugs give evidence that there are various substances which accelerate lung maturation, and as a result, are possibly suitable to prevent the membrane syndrome. In animal experiments aminophyllin promoted the prenatal activity of the surfactant system. It has been clinically proved that short-term application of β-mimetics given in cases of premature labor decreases the number of membrane syndromes [1]. It has also been proved betamimetics promote the secretion of surfactant into the alveoli in animals [6].

On the other hand, however, when long-term application of β-mimetics is used, lung compliations occur more frequently due to the influence of the β-mimetics on lung maturation. This would confirm our observation during the first years after the introduction of long-term β-mimetic therapy, when 3 mature, normal-weight newborns in our department developed a membrane syndrome. 2 of these infants died as a result of this complication. We have investigated in 60 pregnant women with 157 transabdominal amniocenteses whether long-term application of the β-mimetic fenoterol influences the lecithin content of the amniotic fluid [2]. The lecithin values group treated with long-term β-mimetic therapy lie clearly below those of the control group. In figure 1 the median values are shown. The difference in distribution is significant from week 33/0 to 39/6.

Table I. Number of cases with a lecithin content below the critical limit (3 mg/100 ml amniotic fluid) in the group on long-term β-mimetics (157 determinations in 60 pregnant women) and in the control group (211 determinations in 190 pregnant women) according to the week of pregnancy

	Control group			Group on long-term β-mimetics		
	total	<3 mg lecithins 100 ml		total	<3 mg lecithins 100 ml	
	N	n	%	N	n	%
33/0–6	20	10	50	11	10	90
34/0–6	22	11	50	19	16	84
35/0–6	27	10	37	35	26	74
36/0–6	33	7	21	37	26	70
37/0–6	20	2	10	26	19	73
38/0–6	39	2	5	20	8	40
39/0–6	50	5	10	9	5	55

However important the distribution of the lecithin values, it is still more important, in clinical practice, to know how often values that reveal insufficient lung maturity occur in the group on long-term β-mimetics compared to the control group (table I). It is apparent that in the group on long-term β-mimetic therapy at gestational ages between 33/0 and 39/6 weeks, values below the critical limit of 3 mg/100 ml amniotic fluid occur much more frequently than in the control group.

The clinical conclusion from our results is that long-term tocolytic therapy should only be discontinued when the lecithin level in the amniotic fluid lies above the critical limit of 3 mg/100 ml.

The influence of thyroid hormones on lung maturity was established as well: newborns who later on developed a membrane syndrome exhibited low levels of thyroid hormones in the umbilical cord blood [12]. In animal experiments the effect of thyroxin on the function of the lung could be proved [8,15], and in vitro tests on fetal rabbit lung cells showed increased reception of cholin and incorporation into the lecithin after administration of thyroxin [13]. Administration of intraamniotic thyroxin to promote fetal lung maturation in man is being discussed [10]. It was therefore important to measure the concentrations of thyroid hormones in the amniotic fluid to compare them with the lecithin values [3].

From April 1975 to October 1976 we performed clinical amniocenteses in 256 pregnant women at 31/0 weeks of pregnancy. Whereas thyroxin and lecithin levels showed no correlation, the reverse triiodothyronine and lecithin values showed a significant correlation. At the present time we are examining experimentally the possibilities of promoting lung maturation with thyroxin.

Perinatal lung development offers a suitable model for physicochemical, physiological and histological studies. Further investigations will have to find the best way for preventing the membrane syndrome.

References

1 Boog, G.; Ben Brahim, M., and Gandar, R.: Betamimetic drugs and possible prevention of respiratory distress syndrome. Br. J. Obstet. Gynaec. *82:* 285 (1975).
2 Dudenhausen, J.W.; Kynast, G.; Lange-Lindberg, A.-M., and Saling, E.: Influence of long-term beta-mimetic therapy on the lecithin content of amniotic fluid. Gynecol. obstet. Invest. *9:* 205 (1978).
3 Dudenhausen, J.W.; Meinhold, H. und Kynast, G.: Fruchtwasser-Konzentrationen des Thyroxins, des reversen Trijodthyronins und der Lezithine. Fortschr. Med. *97:* 1001 (1979).
4 Farell. P.M. and Zachman, R.D.: Induction of choline phosphotransferase and lecithin synthesis in the fetal lung by corticosteroids. Science *179:* 297 (1973).
5 Fencl, M. De M. and Tulchinsky, D.: Total cortisol in amniotic fluid and fetal lung maturation. New Engl. J. Med. *292:* 133 (1975).
6 Gerner, R.: Untersuchungen zur Beeinflussung der Lezithinsynthese in der fetalen Lunge durch Tokolytika; in Schmidt, Dudenhausen und Saling, Perinatale Medizin, 8. Dt. Kongr. Perinat. Med., Berlin 1977, vol. 7 (Thieme, Stuttgart 1978).
7 Hallmann, M. and Gluck, L.: Development of the fetal lung. J. perinatal Med. *5:* 3 (1977).
8 Hemberger, J.A. and Schanker, L.S.: Effect of thyroxine on the permeability of the neonatal rat lung to drugs. Biol. Neonate *34:* 299 (1978).
9 Kynast, G. and Saling, E.: Routine diagnosis of fetal lung maturity. J. perinatal Med. *2:* 208 (1974).
10 Mashiach, S.; Barkai, G.; Sack, J.; Stern, E.; Goldmann, B.; Brish, M., and Serr, D.M.: Enhancement of fetal lung maturity by intraamniotic administration of thyroid hormone. Am. J. Obstet. Gynec. *130:* 289 (1978).
11 Murphy, B.E.P.: Cortisol and cortisone levels in the cord blood at delivery of infants with and without the respiratory distress syndrome. Am. J. Obstet. Gynec. *119:* 1112 (1974).
12 Redding, R.A. and Pereira, C.: Thyroid function in respiratory distress syndrome (RDS) of the newborn. Pediatrics, Springfield *54:* 423 (1974).
13 Smith, B.T. and Torday, J.S.: Factors affecting lecithin synthesis by fetal lung cells in culture. Pediat. Res. *8:* 848 (1974).

14   Vogel, M.: Morphologische Voraussetzungen für den Einsatz der pränantalen Lungen-
     reifeförderung; in Schmidt, Dudenhausen und Saling, Perinatale Medizin, 8. Dt. Kongr.
     Perinatale Medizin, Berlin 1977, vol. 7 (Thieme, Stuttgart 1978).
15   Wu, B.; Kikkawa, Y.; Orzalesi, M.M.; Motoyama, E.K.; Kaibora, M.; Zigas, C.J., and
     Cook, C.D.: The effect of thyroxine on the maturation of fetal rabbit lungs. Biol.
     Neonate 22: 161 (1973).

PD. Dr. med. Joachim W. Dudenhausen, Abteilung für Geburtsmedizin der Frauenklinik Neuköln, Arbeitsgruppe Perinatale Medizin der Freien Universität Berlin,
Mariendorfer Weg 28, D-1000 Berlin 44

Discussion

Answering a question by *Whittle, Dudenhausen* pointed out that a correlation between
L/S ratio and incidence of respiratory distress syndrome (RDS) in the cases with long-term β-
mimetic therapy was not attempted. *Hallman* remarked that other factors besides lecithin
catabolism or changes in enzymes may affect the L/S ratio. Answering a question by *Mietens,
Dudenhausen* stated that higher cortisol levels in amnion infection may be responsible for
elevated L/S ratios. In answer to *Hallman's* comment, *Dudenhausen* explained that
controlled studies dealing with the effect of β-mimetics on RDS have demonstrated that the
frequency of RDS was lower during short-term therapy and that there may be an important
difference between long-term and short-term β-mimetic therapy.

Prog. Resp. Res., vol. 15, pp. 155–167 (Karger, Basel 1981)

Diagnostic Problems of Surfactant Defects during Birth

K. Diedrich, F. Lehmann, D. Krebs

Departments of Gynecology and Obstetrics I and II, School of Medicine, Lübeck

Introduction

The decision to terminate a pregnancy prematurely is mainly influenced by the possible risk of respiratory distress syndrome (RDS) which is responsible for more perinatal deaths than any other disease [7]. The fact, that the majority of infants with RDS are born prematurely, suggests that incomplete development of the fetal lung is a critical factor. The lungs of these neonates are deficient in surfactant, which causes the alveoli to collapse at the end of expiration [9, 25]. The introduction of methods for the assessment of fetal lung maturity was one of the important advances in the management of high-risk pregnancies and premature infants. Intrauterine fetal respiratory movements result in extrusion of endotracheal fluid which causes the pulmonary surfactant to penetrate the amniotic fluid [1, 15] in which the stage of lung maturity achieved can be determined by simple amniocentesis.

The necessity for such an estimation of the pulmonary surfactant in amniotic fluid arises particularly in high-risk pregnancies including EPH gestosis, Rh incompatibility, placenta previa, placental insufficiency and premature labor [3]. When measuring the phospholipid content in the amniotic fluid, fetal pulmonary maturity can generally be estimated *ante partum*. It was a paper by *Gluck et al.* that prompted a flood of publications dealing with the measurement of surfactant in amniotic fluid. *Gluck and Kulovich* [11] demonstrated that there is an increase in lecithin concentration in the amniotic fluid when fetal lung maturity is reached. Furthermore, this rise also reflects a change in the metabolism of lecithin in the fetal lung which synthesizes a molecule of greater surface activity, dipalmitoyl lecithin.

Table I. Physical methods for the estimation of fetal lung maturity

Method	Apparatus	Test time min	False negative %	False positive %	References
Measurement of[1] the surface tension	Wilhelmy balance	45	9–27	3–5	28 22
Clements test[1]	none	30	18–37	4–19	2, 14, 29
Fluorescence polarization	micro-viscosimeter	20	13	1	6, 26

[1] These methods should be used for screening only.

The pulmonary surfactant is a complex that coats the pulmonary alveoli. It is composed of a variety of substances including proteins and carbohydrates, but consists mainly of lipids [9]. The main compound is a surface-active lecithin containing saturated fatty acids on both the α- and β-carbons. Lecithin makes up nearly 80% of the surfactant phospholipids; other, minor, components are phosphatidylinositol, phosphatidylethanolamine, sphingomyelin and phosphatidylserine. The second major phospholipid is phosphatidylglycerol which makes up more than 10% of surfactant phospholipids in the adult [16].

Methods

Numerous methods for the estimation of pulmonary surfactant are available at the present time. They vary considerably in the work involved as well as in the reliability of their results. In the present report, the various diagnostic methods, both physical and chemical, applied for measuring the surfactant in the amniotic fluid have been reviewed.

Physical Methods

The physical methods, generally used for screening are the measurement of surface tension and the foam test of *Clements et al.* [2] (table I).

Surface tension can be determined by a modification of the Wilhelmy plate method [19, 22, 28]. This method offers the possibility to obtain information on the surfactant within a very short time. A comparison with

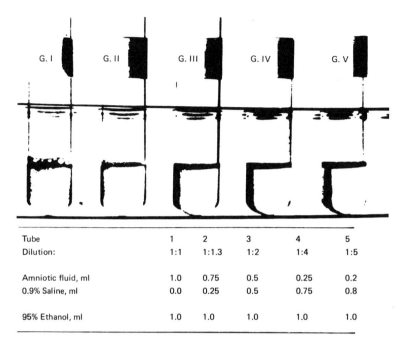

Tube			1	2	3	4	5
Dilution:			1:1	1:1.3	1:2	1:4	1:5
Amniotic fluid, ml			1.0	0.75	0.5	0.25	0.2
0.9% Saline, ml			0.0	0.25	0.5	0.75	0.8
95% Ethanol, ml			1.0	1.0	1.0	1.0	1.0

Fig. 1. The Clements test.

the clinical data shows that there are only few false-positive values but 20% false-negative surface tension values. The negative effect of blood contamination on surface tension is well documented and has been demonstrated for most of the methods evaluating fetal lung maturity.

In 1972 *Clements et al.* [2] reported a simple bedsidetest using a minimum of apparatus and materials with a high predictive value for the occurrence of RDS, when assessed in uncontaminated amniotic fluid. The Clements test is performed as shown on figure 1. It is based on the ability of the pulmonary surfactant to generate stable foam in the presence of ethanol. The results were classified as negative, intermediate or positive. The percentage of false-negative results is 18–37% [14, 29]. False-positive results in this test range from 4 to 19%. The number of false-negative and false-positive results can be decreased when certain details of the technique are fulfilled: all glassware must be free of any kind of detergent, the sample must be free of blood and meconium contamination and uncentrifuged. It is also very important to have an exact water-ethanol mixture.

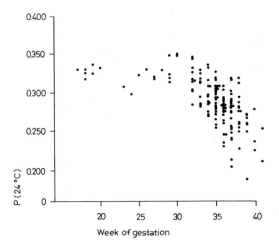

Fig. 2. Fluorescence polarization (P) in the amniotic fluid (n = 147).

Table II. Fluorescence polarization and the incidence of RDS

Fluorescence polarization index	Number of newborns	RDS			No RDS
		severe	moderate	slight	
> 0.310	7	4	2	1	0
0.291–0.310	8	1	3	2	2
≤ 0.290	72			1	71

The Clements test can serve as a screening test for a comparison with other techniques for the determination of surfactant. The experience of most investigators is, that the test is reliable if the results are positive but not if they are negative or doubtful. The Clements test should only be used if none of the other methods to be presented can be carried out. On the other hand, if it is positive, it will be useful in clinics which do not have a specialized laboratory.

Another, very new, physical determination of fetal lung maturity should also be mentioned. The molecular mobility can be determined spectrophotometrically by measuring flucorescence polarization (P) in lipid solutions [6, 26]. This method of determining the microviscosity in lipid solutions can now also be applied for the analysis of amniotic fluid.

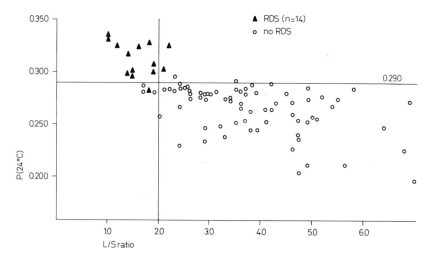

Fig. 3. Incidence of RDS. Comparison between fluorescence polarization and L/S ratio.

The P value is proportional up to the lipid concentration in the amniotic fluid and is in good correlation with fetal lung maturity. The results of fluorescence polarization in samples of amniotic fluid examined in our laboratory are shown in figure 2. The P values show a steady state of 0.350 between weeks 19 and 31 of gestation. There is a steep decline between weeks 31 and 36. 0.290 is the critical P value above which pulmonary maturity cannot be expected and *post partum* RDS is probable. Table II shows that when the P value was above 0.290 a RDS occurred in 13 out of 15 neonates whereas when the P value was less than 0.290, only 1 of 72 neonates developed slight RDS. It is obvious that the prediction of pulmonary maturity is as safe with the method of fluorescence polarization as with the L/S ratio method (fig. 3). The short time needed for the test and also the reliability of its results speak in favor of the method of fluorescence polarization.

Chemical Methods

Chemical analyses of the amniotic fluid constitute another approach to obtain information on the surfactant. The principle of all these analytical tests is the determination of single components of the lipids which are then compared to the clinical data (table III).

Table III. Chemical methods

	Apparatus	Test time h	False negative %	False positive %	References
L/S ratio	densitometer thin layer chromatography	3–4	7–15	1–5	4, 23, 32, 33
Total lipid concentration	densitometer	3–4	9–13	2–12	1, 34
Lecithin concentration					
chromatography	densitometer	4	2–5	1–3	24
enzymatically	photometry	1	8	0	5
Palmitic acid estimation	gas chromatography	2	= 5	= 2	20, 31
Phosphatidyl-glycerol concentration	densitometer	4–5	= 3	0	16

Table IV. The L/S ratio method

Amniotic fluid sample
Centrifugation (500 g)
1 ml supernatant + 2 ml methanol + 2 ml chloroform
Centrifugation (500 g)
Evaporation with N_2
Dissolution in 20 µg chloroform
Thin layer chromatography in chloroform/methanol/ammoniac solution (140/60/10)
Spraying of the plates with bromthymol blue 0.4 g/l
Densitometry

One of the methods for separating the lipids of the amniotic fluid is thin layer chromatography. By this method the L/S ratio can be calculated. Table IV shows the method applied in our laboratory for estimating the L/S ratio [4]. After centrifugation of the sample, the phospholipids are extracted with chloroform and methanol and separated by thin layer chromatography. The stained areas of lecithin and sphingomyelin on the chromatogram are then quantitated by densitometry. According to *Gluck*

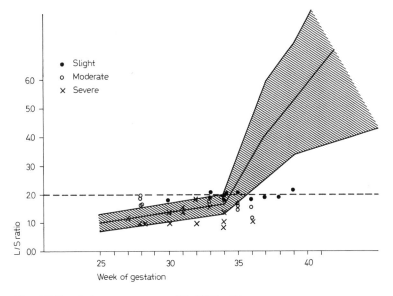

Fig. 4. L/S ratio in pregnancy. n = 421; RDS = 37.

et al. [13] and others RDS has not longer to be expected when the L/S ratio exceeds 2.0.

The results obtained with this method are summarized in figure 4. The L/S ratio is plotted against the times of gestation at which the samples of amniotic fluid were collected. The L/S ratio increases between 31 and 37 weeks of gestation. 421 samples were determined. The incidence of RDS and its graduation in slight, moderate and severe is also shown on this figure. The occurrence of RDS is objectively evaluated by the Hobel-Silverman score [17, 27]. The incidence of false-positive or false-negative results in this test by our method is between 1 and 5%. Other authors have reported about 15% false-negative results [23, 32].

Sources of error mainly result from the instability of the phospholipids in amniotic fluid. Therefore, the sample should be centrifuged as soon as possible. The supernatant should be kept frozen to prevent the enzymes of the amniotic fluid to destroy the lecithin. Furthermore, vaginal samples become easily contaminated with mucus which also contains lipids and will lead to unreliable results. Only a free-flowing fluid should be used when taken from the vagina. Excessive contamination with blood of the mother will give false-positive results as well because of the serum phospholipids. Meconium, if present, will interfere with the extraction of lipids [8]. Also the mode of evaluation of the areas on the chromatogram may lead to false results, especially when the areas are assessed planimetrically. Therefore, densitometric quantification of the areas is recommended.

Gluck et al. [12] placed great emphasis on an intermediate step that consists in the addition of cold acetone to the total lipid extract. It was shown that the resulting precipitate contains the fully saturated surfactant lecithin whereas the other lecithins remain in the supernatant. The value of cold acetone precipitation has been questioned by some investigators [7, 8, 21]. These authors found no differences in fatty acid composition or in the minimal surface tension of the two fractions and therefore omit the acetone precipitation step.

Table III summarizes the other chemical methods for the estimation of the surfactant. Some remarks should be made as concerns the different methods listed. These methods demand various amounts of laboratory work. The percentage of false-negative and false-positive values are noted. Estimation of total lipid concentration has not proved to be of great value. The results obtained by various authors differ considerably [20, 24]. A more promising approach is the assessment of total palmitic acid and lecithin palmitic acid in amniotic fluid, because it correlates precisely with the development of RDS [10, 31]. The analysis of phosphatidylglycerol in amniotic fluid seems to be of value as an additional index of prenatal fetal lung maturity. It may be particulary useful when the specimen is contaminated with blood [16]. Regarding accuracy and technical simplicity, this method offers some advantages over the estimation of the L/S ratio.

Attention, however, has turned to a method which estimates quantitavely one active lipid, as for example lecithin, in the amniotic fluid. The test had to be a simple, rapid and inexpensive, so we concentrated on the main component of the surfactant, lecithin, and developed a specific enzymatic assay for the quantitative determination of lecithin in amniotic fluid [5]. This assay gives absolute concentrations and does not require determination of the L/S ratio.

Lecithin is hydrolyzed by phospholipase C and alkaline phosphatase. Thereafter, the enzymes are deactivated and the choline formed is determined by cholinekinase, pyruvatekinase and lactatedehydrogenase. The single steps are shown in figure 5. Measuring the amount of liberated choline spectrophotometrically, it is possible to determine the absolute lecithin content in the amniotic fluid. The enzymatic determination requires about 45 min.

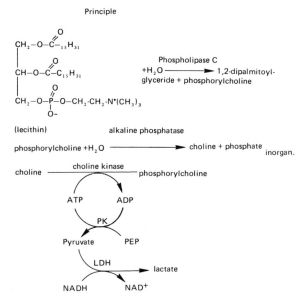

Fig. 5. The enzymatic determination of lecithin.

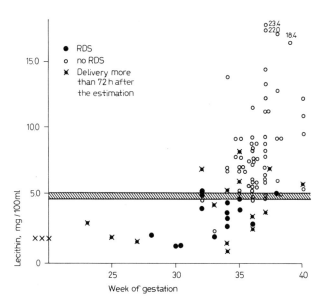

Fig. 6. Quantitative determination of lecithin in the amniotic fluid. n = 99.

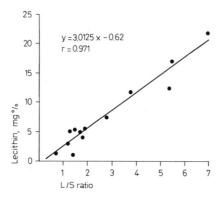

Fig. 7. The correlation between the enzymatic determination of lecithin and the L/S ratio.

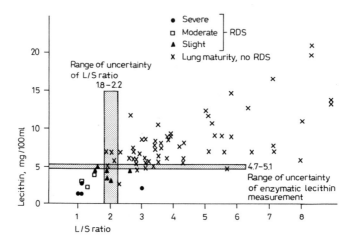

Fig. 8. Incidence of RDS. Comparison of the enzymatic determination of lecithin and the L/S ratio.

The quantitative determination of lecithin in the amiotic fluid by an enzymatic assay is shown in figure 6. As expected, the same increase in lecithin content can be seen as shown with the L/S ratio. RDS is expected in samples with less than 4.7 mg/100 ml lecithin. Lung maturity is expected with lecithin contents of more than 5.1 mg. There is a range of uncertainty between 4.7 and 5.1 mg/100 ml.

Comparison of the quantitative enzymatic lecithin determination and the L/S ratio shows a highly significant correlation between these two methods with a correlation coefficient of 0.971 (p = < 0.001; fig. 7). This enzymatic lecithin determination compares well with the results of the estimation of the L/S ratio, in connection with RDS (fig. 8).

Conclusion

The following methods were presented:

(1) The surface tension test requires a test time of about 45 min. The measurement is easy to perform. The false-negative and false-positive results are relatively high.

(2) The Clements test is easy to perform, needs even less time and has also the disadvantage of a relatively high percentage of false-positive and false-negative values.

(3) Fluorescencepolarization gives exact results within a short time and is well correlated to clinical findings. The main disadvantage is the expensive laboratory equipment.

(4) The L/S ratio is the reference method for every other test. If densitometry is carried out, the equipment is expensive. Thin layer chromatography and planimetry are possible.

(5) Presently, the other chromatographic and gas-chromatographic determinations mentioned here can only be considered as specialized methods of some laboratories. Further investigation will show whether standardization of these tests is possible.

(6) The enzymatic estimation of lecithin gives results in a short time without the need of expensive equipment. The first results are promising. Further application in various laboratories seems to be desirable. Correlation to the fetal outcome is the same as that of the L/S ratio.

Summary:

The risk of RDS has a significant influence on the timing of premature termination of a pregnancy. Numerous methods of physical and chemical analysis of the amniotic fluid for the estimation of fetal lung maturity exhibiting considerable differences in accuracy and time consumption are available today. The various diagnostic methods for measuring amniotic fluid surfactant are reviewed and evaluated. In conclusion, the L/S ratio is the reference method for every other test. The enzymatic determination of lecithin in amniotic fluid offers an easy, highly accurate method which is equivalent to the other mentioned much more time-consuming methods.

References

1   Biezenski, J.J.: Pulmonary surfactant and amniotic fluid lipids; in Obstetrics and gynecology, vol. 5, pp. 71–102 (Appleton-Century-Crofts, New Jersey 1976).

2   Clements, J.A.; Platzker, A.C.G.; Tierney, D.F.; Hobel, C.J.; Creasy, R.K.; Margolis, A.J.; Thibeault, D.W.; Tooley, W.H., and Oh, W.: Assessment of the risk of the respiratory distress syndrome by a rapid test for surfactant in amniotic fluid. New Engl. J. Med. *286:* 1077–1081 (1972).

3   Diedrich, K.: Indikationen zur Bestimmung der fetalen Lungenreife. Gynäk. Prax. *3:* 17–21 (1979).

4   Diedrich, K.; Stefan, M., and Krebs, D.: The effect of betamethasone therapy on the L/S ratio in amniotic fluid. J. perinatal. Med. *6:* 22–27 (1978).

5   Diedrich, K.; Hepp, S.; Welker, H.; Krebs, D.; Beutler, H-O. und Michal, G.: Die enzymatische Lecithinbestimmung im Fruchtwasser zur Beurteilung der fetalen Lungenreife. Geburtsh. Frauenheilk. *39:* 238–246 (1979).

6   Diedrich, K.; Roth, G. und Krebs, D.: Die Messung der Fluoreszenzpolarisation im Fruchtwasser zur Bestimmung der fetalen Lungenreife. Z. Geburtsh. Perinat. *182:* 405–409 (1978).

7   Dunn, L.J. and Bhatnagar, A.S.: Use of lecithin/sphingomyelin ratio in the management of the problem obstetric patient. Am. J. Obstet. Gynec. *115:* 687–692 (1973).

8   Forman, D.T.: The lecithin to sphingomyelin ratio in amniotic fluid and its predictive value for fetal lung maturity (respiratory distress syndrome); in Amniotic fluid, pp. 247–258 (Wiley, New York 1974).

9   Gluck, L.: Surfactant: 1972. Pediat. Clins. N. Am. *19:* 325–331 (1972).

10 Gluck, L.; Kulovich, M.V.; Borer, R.C.; Brenner, P.H.; Anderson, G.G., and Spellacy, W.N.: Diagnosis of the respiratory distress syndrome by amnicentesis. Am. J. Obstet. Gynec. *109:* 440–445 (1971).

11 Gluck, L. and Kulovich, M.V.: Lecithin/sphingomyelin ratios in amniotic fluid in normal and abnormal pregnancies. Am. J. Obstet. Gynec. *115:* 539–546 (1974).

12 Gluck, L.; Kulovich, M.V.; Borer, R.C., and Keidel, N.W.: The interpretation and significance of the lecithin/sphingomyelin ratio in amniotic fluid. Am. J. Obstet. Gynec. *120:* 142–155 (1974).

13 Gluck, L.; Sribney, M.; and Kulovich, M.V.: The biochemical development of surface activity in mammalian lung. Pediat. Res. *1:* 247–265 (1967).

14 Goldstein, A.S.; Fukunaga, K.; Malachowski, N., and Johnson, J.D.: A comparison of the lecithin/sphingomyelin ratio and shake test for estimating fetal pulmonary maturity. Am. J. Obstet. Gynec. *118:* 1132–1138 (1974).

15 Hallman, M. and Gluck, L.: Development of the fetal lung. J. perinatal. Med. *5:* 3–31 (1977).

16 Hallman, M.; Kulovich, M.V.; Kirkpatrick, E.; Sugarman, R.G., and Gluck, L.: Phosphatidylinositol and phosphatidylglycerol in amniotic fluid: indices of lung maturity. Am. J. Obstet. Gynec. *125:* 613–617 (1976).

17 Hobel, C.J.; Oh, W.; Hyvarinen, M.A., and Ehrenberg, A.: Early versus late treatment of neonatal acidosis in low birth weight infants: relation to respiratory distress syndrome. J. Pediat. *81:* 1178–1189 (1972).

18 Kling, O.R.; Crosby, W.M., and Merrill, J.A.: Amniotic fluid correlates of fetal maturity and perinatal outcome. Gynecol. Invest. *4:* 38–49 (1973).

19 Masson, D.; Diedrich, K.; Rehm, G.; Stefan, M. und Schultze-Mosgau, H.: Die Messung der Oberflächenspannung im Fruchtwasser als einfache Methode zur Bestimmung der fetalen Lungenreife. Geburtsh. Frauenheilk. *37:* 57–63 (1977).

20 Moore, R.A.; O'Neil, K.T.J.; Cooke, R.J., and MacLennan, A.H.: Palmitic acid and lecithin measurements in amniotic fluid. Br. J. Obstet. Gynaec. *82:* 194–198 (1975).

21 Morrison, J.C.; Wiser, W.L.; Arnold, J.W.; Whybren, W.D.; Morrison, D.L., Fish, S.A., and Bucovaz, E.T.: Modification of the lecithin/sphingomyelin ratio for fetal development. Am. J. Obstet. Gynec. *120:* 1087–1093 (1974).

22 Müller-Tyl, E.; Lempert, J.; Steinbereithner, K., and Benzer, H.: Surface properties of the amniotic fluid in normal pregnancy. Am. J. Obstet. Gynec. *122:* 295–300 (1974).

23 Nelson, G.H.: Relationship between amniotic fluid lecithin concentration and respiratory distress syndrome. Am. J. Obstet. Gynec. *112:* 827–833 (1972).

24 Ramzin, M.S.: Fruchtwasseruntersuchungen am Ende der Schwangerschaft: Phospholipide in der Spätschwangerschaft. Gynäkologe *6:* 206–213 (1973).

25 Scarpelli, E.M.: Physiologie und Pathologie des pulmonalen oberflächenaktiven Systems. Triangel *10:* 47–56 (1971).

26 Shinitzky, M.; Goldfisher, A.; Bruck, A.; Goldman, B.; Stern, E; Barkai, G.; Mashiach, S., and Serr, D.M.: A new method for assessment of fetal lung maturity. Br. J. Obstet. Gynaec. *83:* 838–844 (1976).

27 Silverman, W.A. and Anderson, D.H.: Obstructive respiratory sign, death rate and necropsy findings among mature infants. Pediatrics, Springfield *17:*1 (1956).

28 Tiwary, C. and Goldkrand, J.W.: Assessment of fetal pulmonic maturity by measurement of the surface tension of amniotic fluid lipid extract. Obstet. Gynecol., N.Y. *48:* 191–194 (1976).

29 Wagstaff, T.I.: The estimation of pulmonary surfactant in amniotic fluid; in Amniotic fluid – Research and clinical application, pp. 347–391 (Excerpta Medica, New York 1978).

30 Wagstaff, T.I.; Whyley, G.A., and Freedman, G.: Factors influencing the measurement of the lecithin/sphingomyelin ratio in amniotic fluid. J. Obstet. Gynaec. Br. Commonw. *81:* 264–271 (1974).

31 Warren, C.; Holton, J.B., and Allen, J.T.: Assessment of fetal lung maturity by estimation of amniotic fluid palmitic acid. Br. med. J. *i:* 94–101 (1974).

32 Lorenz, U.; Rüttgers, H.; Fromme, M., and Kubli, F.: Significance of amniotic fluid phospholipid determination for the prediction of neonatal respiratory syndrome. Z.Geburtsh.Perinat. *179:* 101–111 (1975).

33 Donald, I.R.; Freeman, R.K.; Goebelsmann, U.; Chan, W.H., and Nakamura, Clinical experience with the amniotic fluid lecithin/sphingomyelin ratio. Am.J.Obstet.Gynecol. *115:* 547–552 (1973).

34 Caspi, E.; Schreyer, P., and Tamir, I.: The amniotic fluid foam test. L/S ratio, and total phospholipids in the evaluation of fetal lung maturity. Am.J.Obstet.Gynecol. *122:* 323–326 (1975).

Dr. K. Diedrich, Department of Gynecology and Obstetrics I and II, School of Medicine, Ratzeburger Allee 160, D-2400 Lübeck (FRG)

Prog. Resp. Res., vol. 15, pp. 168–176 (Karger, Basel 1981)

Estimation of Fetal Lung Maturity by Means of Two-Dimensional Thin Layer Chromatography and Microviscosimetry of the Aminiotic Fluid

U. Lorenz, H. Rüttgers, D. Kreml and F. Kubli

Universitäts-Frauenklinik, Heidelberg

Determination of fetal lung maturity by means of amniotic fluid phospholipid measurements has been a routine method for years [1]. Two new methods were described during the past 2 years: the determination of amniotic fluid microviscosity [5] and measurement of minor phospholipid components as phosphatidylserine, phosphatidyl-ethanolamine, phosphatidyl-inositol, phosphatidyl-glycerol (PG) in the amniotic fluid [2] and tracheal secretion [4]. Especially PG seems to play an important role in the composition of surfactant and its restitution in the course of treatment of RDS.

We compared three methods of amniotic fluid phospholipid assessment with regard to prediction of RDS: (1) one-dimensional thin layer chromatography (TLC) of amniotic fluid extracts with planimetric measurement of the lecithin-spingomyelin (L/S) ratio; (2) microviscosimetry of the amniotic fluid (P value), and (3) two-dimensional TLC of amniotic fluid extracts with densitometric measurement of the L/S ratio and PG concentration.

222 amniotic fluid samples of 198 patients were investigated by methods 1 and 2; in 85 of these samples method 3 was also performed. 99 samples (including 13 with subsequent respiratory distress syndrome; RDS) were obtained within 72 before delivery, thus allowing a comparison between the phospholipid results in amniotic fluid and the status of the newborn in respect to RDS.

L/S (one-dimensional TLC)

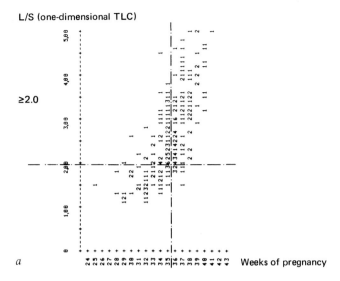

a   Weeks of pregnancy

P value

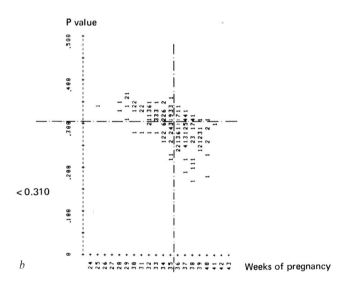

b   Weeks of pregnancy

Fig. 1. One-dimensional TLC of amniotic fluid extracts. Numbers indicate the frequency of identical measurements. *a* L/S ratio versus weeks of pregnancy. *b* Microviscosimetry of amniotic fluid (P value).

L/S (one-dimensional TLC)

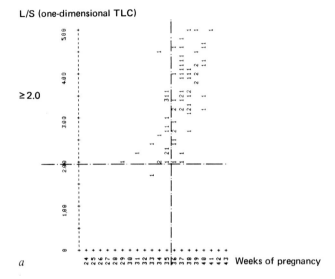

a

L/S (one-dimensional TLC)

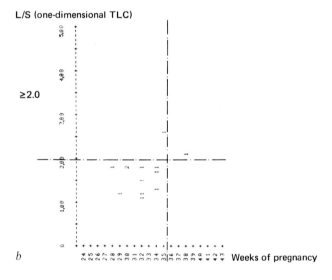

b

Fig. 2. L/S ratios obtained by one-dimensional TLC in newborns without (a) or with (b) RDS. Interval between amniocentesis and delivery ≤ 72 h.

Results

Before 35 weeks of gestation low L/S ratios are predominant, after 35 weeks, a sharp rise towards values > 2 takes place. The P value drops simultaneously, indicating an enhancement of lipid substances in the amniotic fluid (fig. 1). In amniotic fluid samples obtained within 72 h before birth, L/S ratios in the group without subsequent RDS are all but 1 in the range ≥ 2. 11 of 13 cases with RDS show values < 2 (fig. 2). The P values in the group without subsequent RDS are much more scattered above and below 0.310. 3 of 13 cases with subsequent RDS show values < 0.310, falsely indicating lung maturity (fig. 3).

The distribution of values determined by two-dimensional TLC is similar to that obtained by one-dimensional TLC (fig. 4). PG concentration in amniotic fluid is < 20 µg% before 35 weeks of pregnancy and rises after 35 weeks. In no case with RDS PG could be detected in the amniotic fluid. But in cases without RDS also very low PG concentrations are seen, even after 35 weeks of gestation (fig. 5).

For the interpretation of the results, three probability figures (table I) have to be taken into account: (1) probability of correct prediction of RDS; (2) probability of occurrence of RDS, and (3) probability of non-occurrence of RDS. An ideal method for determining fetal lung maturity would show 100% for all three probability figures. For the parameters investigated by us, the figures are shown in table II. The best results are obtained by one- or two-dimensional determination of the L/S ratio. Slightly worse are the results for the determination of P; with values > 0.310, other methods have to be used to define the risk of occurrence of RDS. The determination of the PG fraction alone is not sufficient for RDS prediction, because the rise of PG seems to be related too closely to the duration of pregnancy and there are too many cases beyond 35 weeks which do not show measurable amounts of PG. In cases with L/S ratios close to 2, PG determination might be helpful to calculate the risk of RDS, which is very unlikely if PG is detectable. Also in cases with diabetes, where the L/S ratio alone is not sufficient for the prediction of lung maturity, PG determination might serve as an additional safety factor. Recently, we could show that after *ante partum* administration of betamethasone a significant rise of PG in rat lung homogenates, takes place [3]. If this finding can be reproduced in human amniotic fluid, the determination of PG will become a relevant chemical test for the course and the effect of *ante partum* glucocorticoid treatment.

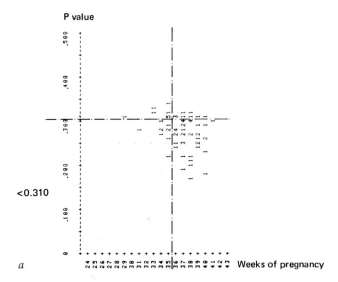

a

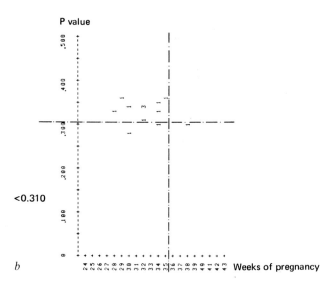

b

Fig. 3. Microviscosimetry in newborns without (a) or with (b) RDS. Interval between amniocentesis and delivery ≤ 72 h.

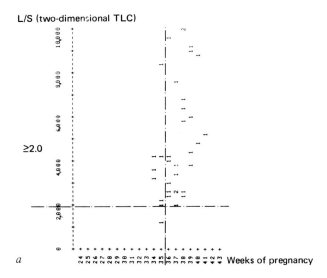

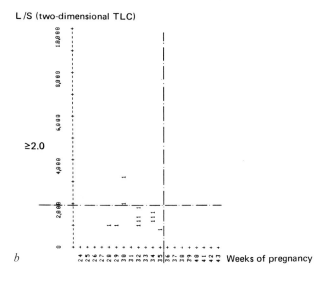

Fig. 4. L/S ratios obtained by two-dimensional TLC in newborns without (a) or with (b) RDS.

PG/100 ml amniotic fluid, µg

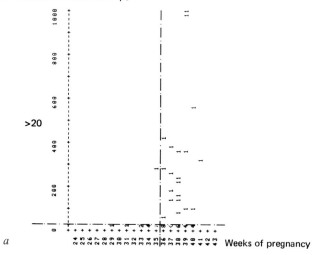

a   Weeks of pregnancy

PG/100 ml amniotic fluid, µg

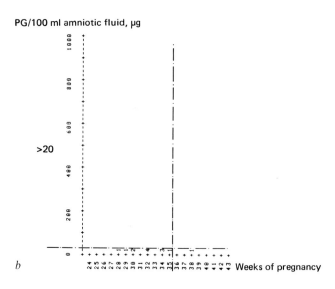

b   Weeks of pregnancy

Fig. 5. PG concentration in amniotic fluid extract in newborns without (a) or with (b) RDS.

Table I. Formulas for the evaluation of the predictive reliability of phospholipid parameters

Probability of correct prediction of RDS:
$$\frac{n \cdot \text{RDS with L/S ratio} < 2}{n \text{ RDS}} \cdot 100$$

Probability of occurrence of RDS
$$\frac{n \text{ L/S ratio values} < 2 \text{ with RDS}}{n \text{ L/S ratio values} < 2 \text{ total}} \cdot 100$$

Probability of non-occurrence of RDS
$$\frac{n \text{ L/S ratio values} \geqslant 2 \text{ with RDS}}{n \text{ L/S ratio values} \geqslant 2 \text{ total}} \cdot 100$$

analogous for	threshold value
P value	0.310
L/S ratio (Two-dimens.)	2.0
PG concentration	20 µg%
PI/PS ratio	0.9

Table II. Probability figures for the phospholipid parameters described in figures 1–5

	L/S ratio	P value	L/S ratio (2-dim)	PG conc.	PI/PS ratio
Probability of correct prediction of RDS, %	85	77	85	100	63
Probability of occurrence of RDS, %	92	34	92	45	50
Probability of non-occurrence of RDS, %	97	96	94	100	88

References

1   Hallman, M., and Gluck, L.: Development of the fetal lung. J. perinat. Med. *5*: 3: (1977).

2   Hallman, M.; Kulovich, M.; Kirkpatrick, E.; Sugarman, R., and Gluck, L.: Phosphatidylinositol and phospatidylglycerol in amniotic fluid: indices of lung maturity. Am. J. Obstet. Gynec. *125*: 613 (1976).

3   Lorenz, U.; Kaufmann, M.; Rüttgers, H.; Paweletz, N.; Heberling, D., and Kubli, F.: The effect of glucocorticoids on the development and maturation of the fetal rat lung. IXth Wld. Congr. Gynec. Obstet., Tokyo 1979.

4   Obladen, M.: Tracheale Phospholipid-Zusammensetzung und Atemnotsyndrom des Neugeborenen; Habilitationsschrift Heidelberg (1977).

5   Shinitzky, M.; Goldfisher, A.; Bruck, A.; Goldman, B.; Stern, E.; Barkai, G.; Mashiah, S., and Serr, D. M.: A new method for assessment of fetal lung maturity Br. J. Obstet. Gynaec. *83*: 838 (1976).

Dr. U. Lorenz, Universitäts-Frauenklinik, Vosstrasse 9, D-6900 Heidelberg (FRG)

Discussion

Answering a question by *Tiebes, Lorenz* explained the microviscosity method. It is an indirect determination of viscosity using a fluorescent material attached to the lipids of the amniotic fluid which depolarizes light. There is no comparison between this method and other viscosimetric methods. Answering a question by *Clements,* he said that there is no proof until now that this additional method will improve the prediction of risk cases for RDS. *Enhorning* comments that any of these data, e.g. L/S ratio or shake test, may be a parameter of gestational age, which is, as generally known, sometimes difficult to assess. These tests, are, therefore, in some cases unspecific for lung maturity. This comment was agreed and stated that the more unspecific tests dealing with crude lipids in amniotic fluid are not so good as the estimation of saturated lecithin, which on the other hand, is unfortunately very complicated. *Clements* pointed out that the foam test was simple and at least allowed semiquantitative determination of surface-active material. There is a good correlation between this method and the RDS risk.

Prog. Resp. Res., vol. 15, pp. 177–187 (Karger, Basel 1981)

Multilamellar Liposome Preparation and Surface Activity of Synthetic Phospholipids

M. Obladen, I. Klatt and M. Bartholomé

Universitäts-Kinderklinik, Heidelberg, and Zentrallabor der Universitäts-Kliniken, Göttingen

Introduction

Phospholipids have been considered the main functional compound of surfactant since *Klaus et al.* [20] found their high concentration in beef lung. The required property, lowering of the alveolar surface tension during expiration, has been attributed to disaturated lecithin, especially to dipalmitoyl phosphatidylcholine (DPPC), which accounts for 73–86% of surfactant phospholipids [9, 33]. Its molecules, arranged as a monomolecular film at the air-water interphase, are believed to stabilize on compression, counteracting the tendency of the alveolar wall to collapse during expiration. However, there is more to surfactant function than sole lowering of the surface tension. As worked out by *Clements* [6], a substance acting as surfactant must be *released* from the cell, *adsorb* from the subphase at physiologic temperature, *spread* to a monolayer and *lower* the surface tension on compression. Active adsorption requires high molecular mobility which is present in phospholipids after their transition from the gel state to the liquid crystal state. DPPC alone has a phase transition temperature of 41 °C [1, 35], will be in the crystalline state at room temperature and is not likely to spread at 37 °C.

Recently, *Hallman et al.* [13] showed total absence of phosphatidylglycerol (PG) in the lung of infants with respiratory distress syndrome (RDS), a finding confirmed by *Cunningham et al.* [7], and our own studies [25]. PG is a polar phospholipid, has a negative charge and is characterized by a low transition temperature of 25 °C [3]. The substantial physical differences between DPPC and dipalmitoyl phosphatidylglycerol (DPPG) led us to study the interaction of these phospholipids by surface activity measurements with a modified Wilhelmy balance.

Materials and Methods

Materials

Cholesterol (99%) and DPPC (98%) were purchased from Sigma, St. Louis, Mo. A mixture of DPPC (99%) and DPPG (82%) 9:1 by weight was a gift from Nattermann, Köln (FRG). Chloroform, analytical grade, was from Merck, Darmstadt, FRG, isotonic NaCl solution was purchased from Fresenius, Bad Homburg (FRG), and proved to be free of surface-active contamination.

Methods

Preparation of Crude Liposomes. Manually shaken liposomes were prepared according to *Bangham et al.*[2] as follows: 10–50 mg phospholipid was dissolved in chloroform and dried to a thin film with a rotary evaporator at 37 °C. The 100-ml round flask contained 20–30 glass spheres of 3 mm in diameter throughout the procedure. The dry film was taken up in 0.9% NaCl solution by gentle shaking of the glass spheres at room temperature. With this procedure, large multilamellar liposomes with monomolecular layers are obtained.

Liposome Fractionation According to Gregoriadis et al. *[11].* 50 mg DPPC/DPPG was dissolved in 5 ml chloroform and dried to a thin film in a 25-ml conical flask containing 3 glass spheres. The dry film was taken up using a vortex mixer. To prepare liposomes of defined size, sonication was applied 4 times for 1 min with the 18-cm titanum probe of a Branson sonifier cell disrupter B 15 with a pulse generator, 30% pulse, 40 W, and defined temperature of 37 and 42 °C. This preparation yields smaller liposomes with bimolecular layers.

Phase Contrast Microscopy of liposomal preparations. It was performed on a Diavert Leitz microscope with a phase contrast objective Phaco L 32/04.

Dynamic Surface Tension Measurements. They were performed with a modified Wilhelmy balance designed according to *Clements* [5], *King and Clements* [18] and *Goerke* [10] equipped with a large Langmuir trough with adjustable temperature. The compressible surface of the Teflon-coated Langmuir trough was 392 cm², subphase volume was 200 ml, compression of the film was performed with a movable Teflon barrier at constant speed; one compression-reexpansion cycle lasted for 4 min. In order to approach physiological conditions, only adsorbed films were studied on a subphase of 0.9% NaCl and at 37 °C. Phospholipid samples of 1 mg were injected into the subphase resulting in a potential film density of 4.89 Å²/molecule; 10 min were allowed to adsorb from the subphase before compression was started.

Evaluation of Hysteresis. Each cycle was calibrated to 73 mN/m with pure NaCl, γ_{max} was the maximal surface tension of an uncompressed film (mN/m), γ_{min} was the minimal surface tension of a film compressed to 20% of initial area. The stability index (SI) was calculated according to *Clements* [5] as

$$SI = 2 \; \frac{\gamma_{max} - \gamma_{min}}{\gamma_{max} + \gamma_{min}} \; .$$

Compliance. To assess intrapulmonary surface activity, a model for surfactant deficiency was used: pressure/volume diagrams of isolated rabbit lungs were recorded during

Table I. Surface activity of different preparations of DPPC/DPPG 9:1

Preparation	n	γmin, mN/m	SI
(1) Dry crystal	11	47.9 ± 21.3	0.48 ± 0.51
(2) Mechanical suspension	10	69.7 ± 3.1	0.04 ± 0.04
(3) Sonicated suspension	20	57.5 ± 21.7	0.31 ± 0.57
(4) Crude multilamellar liposomes [2]	26	3.7 ± 6.8	1.83 ± 0.29
(5) Sonicated liposomes	15	36.1 ± 25.9	0.82 ± 0.74
(6) Fractionated bilayered liposomes [11]	8	67.3 ± 6.0	0.10 ± 0.12
(7) Fractionated liposomes (sonicated with pulse up to 37 °C)	5	23.8 ± 17.4	1.06 ± 0.50
(8) Fractionated liposomes (continuous sonication at 37 °C)	4	2.3 ± 0.1	1.86 ± 0.01
(9) Fractionated liposomes (sonicated at 40–42 °C)	8	2.0 ± 0.5	1.89 ± 0.03

High surface activities are obtained by crude (monolayered) multilamellar liposomes (line 4) and by fractionated (bilayered) liposomes sonicated above the transition temperature.

ventilation with a bulb syringe similar to the method of *Gribetz et al.* [12]. The system was connected to a pressure gauge transducer and a piston potentiometer and ligated into the trachea with a 3.5-mm Portex tube. Inflation and deflation was done at constant speed. After the first measurement, the lung was washed 20 times with 0.9% NaCl portions of tidal volume in order to remove significant amounts of surfactant as white opalescent solution. Repeated pressure/volume recording showed reduced compliance. Then, the phospholipid preparation was instilled into the tracheal tube, the lung inflated several times and incubated at 37 °C for 30 min. A final compliance measurement was performed afterwards as described.

Results

Table I shows the surface activity of different physical preparations of DPPC/DPPG, 9:1. Dry crystals and mechanical suspensions have no surface activity. Sonicated suspensions change from a milky appearance to a clear solution but likewise have minimal surface tension above 50 mN/m. Crude multilamellar liposomes, however, display high surface activity which can be destroyed by sonication for 10 sec to 1 min. The initial liposomal fraction according to *Gregoriadis et al.* [11] on the other hand, is completely inactive, its morphology reveals no structures. If it becomes transformed to monolayers or concentric bimolecular multilamellated liposomes by sonication, they have high surface activity and γ_{min} close to zero.

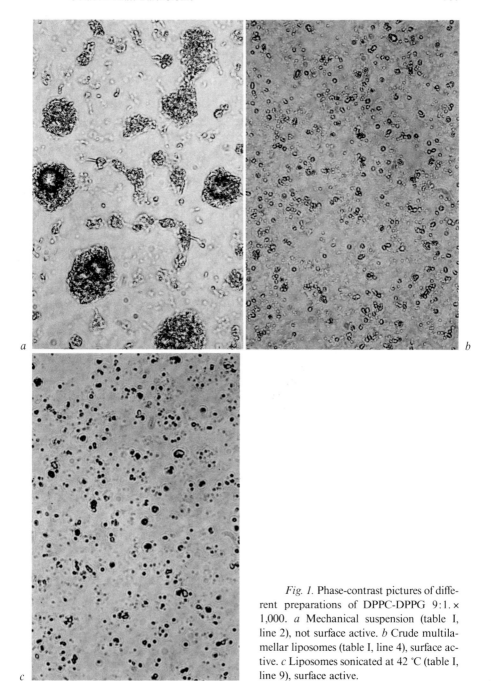

Fig. 1. Phase-contrast pictures of different preparations of DPPC-DPPG 9:1.× 1,000. *a* Mechanical suspension (table I, line 2), not surface active. *b* Crude multilamellar liposomes (table I, line 4), surface active. *c* Liposomes sonicated at 42 °C (table I, line 9), surface active.

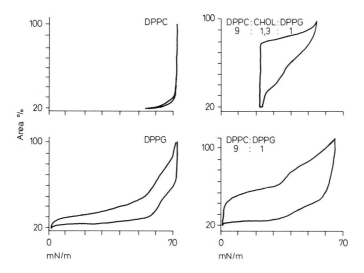

Fig. 2. Surface tension area diagrams of different synthetic phospholipids in a preparation of crude liposomes. 1 mg/200 ml subphase (0.9% NaCl); adsorption time 10 min at 37 °C. At the right of each diagram: calibration line registered before application of sample.

Figure 1 demonstrates phase contrast pictures of fractions 2, 4, and 8 shown in table I. In all fractions with spreading- and surface-tension-lowering ability, multilamellated concentric structures of 200–600 Å are observed. The inactive samples are characterized by amorphous agglomerates.

In figure 2 some typical compression-reexpansion loops of liposomal preparations are reproduced, all of which had been actively adsorbed to the surface after injection into the saline subphase. For DPPC, the minimal surface tension possible cannot be recorded after 10 min adsorption: this lipid adsorbs very slowly at 37 °C and does not reach γ_{min} below 10 mN/m within 90 min. DPPG, however, adsorbs to an active film with γ_{min} of 1.5 mN/m few minutes after administration. The synthetic mixture of DPPC/DPPG, 9:1, is highly surface-active, spreads to the surface within 1 min and has a γ_{min} of 1 mN/m and an SI of 1.93. If in a further approach to mimic natural surfactant composition, cholesterol is added to DPPC/DPPG, surface activity is diminshed and surface tension cannot be decreased below 22.5 mN/m even with higher film concentrations.

The results of a series of measurements in four synthetic phospholipid samples are represented in table II. Again, DPPC shows little surface

Table II. Surface activities of different synthetic phospholipids at 37 °C

Lipid	n	Å²/mol	γmax mN/m	γmin mN/m	SI
DPPC	14	2.45–4.89	70.9 ± 2.0	29.5 ± 20.7	0.95 ± 0.60
DPPG	8	4.89	59.8 ± 5.4	2.6 ± 1.5	1.83 ± 0.09
DPPC/DPPG 9:1	26	4.89	70.4 ± 2.6	3.7 ± 6.8	1.83 ± 0.29
DPPC-cholesterol-DPPG 9:1, 3:1	6	4.89	72.8 ± 0.3	59.1 ± 18.7	0.25 ± 0.40

Active adsorption from the subphase and extensive lowering of the surface tension are the characteristics of DPPG and DPPC : DPPG, 9:1

activity, the high SD for this lipid is due to the subphase temperature of 37 °C, at which it exhibits some slow adsorption leading to a variable degree of film function after 10 min. The high surface activity of DPPG and DPPC/DPPG, 9:1, is confirmed by this series.

The inflation-deflation curve of the isolated rabbit lung shown in figure 3 gives an idea of the possible activity of surface-active phospholipids *in vivo.* The compliance of the lung has been reduced to half its initial value by extensive alveolar lavage removing much of the natural surfactant. After injection of 11 mg DPPC/DPPG 9:1 per kilogram body weight into the tracheal tube, the deflation limb of the pressure/volume diagram closely resembles the initial curve.

Discussion

53% of surfactant dry weight is of lipid origin. 83% of the lipids are phospholipids and 10–16% cholesterol [9, 30]. Surface activity has been ascribed to fatty acids [8], sphingomyelin [19], lecithin [DPPC; 4, 23], PG [15, 16], phosphatidylinositol [14, 24] and cholesterol [31]. Our data do not support the speculation that DPPC alone is responsible for the extensive surface-tension-lowering ability of surfactant, as it does not spread to the surface at physiologic temperatures. Phosphatidylglycerol is the second major surfactant phospholipid and makes up to 10% of phospholipid phosphorus [29, 32]. Our studies show, that the surface-tension-lowering

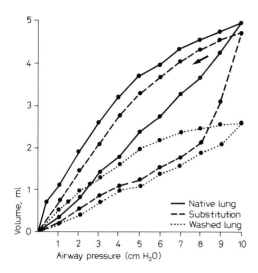

Volume, ml

Airway pressure (cm H₂O)

— Native lung
-- Substitution
···· Washed lung

Fig. 3. Pressure/volume diagrams of isolated rabbit lung before and after lung wash with 78 ml/kg 0.9% NaC1, and after intratracheal application of 11 mg/kg DPPC:DPPG 9:1 in a preparation of crude liposomes. Deflation limb (arrow) after substitution resembles initial curve.

capacity of DPPG is equivalent to DPPC but that it is far more active in its spreading behavior. As PG is totally absent in RDS, disturbed phospholipid interaction might be an important pathogenetic factor and not only diminished lecithin. The enhancement of film spreading suggested by our results might be the modifying function of PG postulated by *Hallman and Gluck* [15]. It is probably due to the negative charge of PG as in vesicles containing both lecithin and PG the charged compound has a preference for the outside of the bilayer [22]. Negatively charged liposomes are characterized by specific proteinphospholipid interactions [17].

Admixture of cholesterol prevents γ_{min} below 20 mN/m [31], its condensing effect impairs fatty acid motility [36]. Cholesterol function in normal respiration is poorly understood.

In order to mimic surfactant functions *in vitro,* it is not enough to provide a phospholipid mixture adapted to the biochemical pattern found *in vivo.* In addition to the chemical properties, the technique of preparation greatly influences the ability of a potential synthetic substitute to spread to an active film. Suspended amorphous material does not reveal any surface activity. Ultrasound creates stable, optically clear solutions consisting of small unilamellar microvesicles [2, 27] without surface activity. Multilamellar liposomes, however, spread actively and fast to the surface and display high surface activity. As it is not easy to prepare reproducible liposome suspensions of uniform size and structure, this finding needs further confirmation. *Pattus et al.* [28] found transformation of closed bilayers into lipid monolayers when spreading liposomes at the air-water interface at zero surface pressure.

Again, as for the coincidence of PG function in a synthetic film and absence of PG in RDS, a striking coincidence of biological and physico-chemical characteristics is present: multilamellar lipsomes prepared *in vitro* from synthetic phospholipids morphologically resemble very much the lamellar inclusion body of the type II cell, the site of surfactant storage and secretion *in vivo* [21, 33, 34]. If phospholipid spreading to active monolayers proves to be a function of multilamellar *structure,* as our data indicate, the therapeutic application of suitable liposomal preparations might offer a possibility to administer synthetic phospholipids to the lung in a surface-active form [26]. Our results showing the high physicochem-ical diversity of chemically identical liposomes prepared in different sizes and at different temperatures support this speculation. Further studies should be focused on the relation of liposome type and surface activity in order to find out appropriate preparative conditions for therapeutic trials.

Summary

The surface activity of synthetic DPPC, and a mixture of DPPC DPPG, 9:1m, was studied for different preparations. Dry substance, mechanical suspensions and sonicated solutions had no surface activity. DPPC did not adsorb to the surface at 37 °C. Liposomal preparations of DPPG and DPPC/DPPG 9:1 show fast adsorption from the subphase at 37 °C and lowered the surface tension to 2.6 and 3.7 mN/m on compression in the Wilhelmy balance. Active samples showed multilayered liposomal structures by phase contrast microscopy and were able to restitute the compliance of a washed rabbit lung. We speculate, that phospholipid spreading to active monolayers is a function of multilamellar structures.

References

1 Albon, N. and Sturtevant, J. M.: Nature of the gel to liquid crystal transition of synthetic phosphatidylcholines. Proc. natn. Acad. Sci. USA *75:* 2258–2260 (1978).

2 Bangham, A.D.; Hill, M.W., and Miller, N.G.A.: Preparation and use of liposomes as models of biological membranes in Korn, Methods in membrane biology, vol. 1, pp. 1–68 (Plenum Press, New York 1974).

3   Bevers, E.M.; Op Den Kamp, J.A.F.; and Van Deenen, L.L.M.: Physico-chemical properties of phosphatidylglycerol in membranes of *Acholeplasma laidlawii.* Eur. J. Biochem. *84:* 35–42 (1978).

4   Brumley, G.W.; Hodson, W.A., and Avery, M.E.: Lung phospholipids and surface tension correlations in infants with and without hyaline membrane disease and in adults. Pediatrics, Springfield *40:* 13 (1967).

5   Clements, J.A.: Surface tension of lung extracts. Proc. Soc. exp. Biol. Med. *95:* 170 (1957).

6   Clements, J.A.: Functions of the alveolar lining. Am. Rev. resp. Dis. *115:* 67–71 (1977).

7   Cunningham, M.D.; Greene, J.M., and Thompson, S.A.: Antenatal reduction of surfactant phosphatidylglycerol (PG) in infants of diabetic mothers (IDM) with respiratory distress (RD). Pediat. Res. *10:* 460 (1976).

8   Friedman, Z. and Rosenberg, A.: Abnormal lung surfactant related to essential fatty acid deficiency in a neonate. Pediatrics, Springfield, *63:* 855–859 (1979).

9   Frosolono, M.F.; Charms, B.L.; Pawlowski, R., and Slivka, S.: Isolation, characterization, and surface chemistry of a surface-active fraction from dog lung. J. Lipid Res. *11:* 439 (1970).

10   Goerke, J.: Lung surfactant. Biochim. biophys. Acta *344:* 241 (1974).

11   Gregoriadis, G.; Leathwood, P.D., and Ryman, B.E.: Enzyme entrapment in liposomes. Febs Lett. *14:* 95–99 (1971).

12   Gribetz, I.; Frank, N.R., and Avery, M.E.: Static volume-pressure relations of excised lungs of infants with hyaline membrane disease, newborn and still-born infants. J. clin. Invest. *38:* 2168–2175 (1959).

13   Hallman, M.; Feldman, B., and Gluck, L.: The absence of phosphatidylglycerol in surfactant. Pediat. Res. *9:* 396 (1975).

14   Hallman, M.; Kulovich, M.; Kirkpatrick, E., and Gluck, L.: Phosphatidylglycerol and phosphatidylinositol in amniotic fluid. Indices of lung maturity. Am. J. Obstet. Gynecol. *125:* 613–617 (1976).

15   Hallman, M. and Gluck, L.: Phosphatidylglycerol in lung surfactant. III. Possible modifier of surfactant function. J. lipid. Res.

16   Henderson, R.F.; Waide, J.J., and Pfleger, R.C.: Methods for determining the fraction of pulmonary surfactant lipid removed from the lung of beagle dogs by lavage. Archs. int. Physiol. Biochim. *82:* 259–272 (1974).

17   Kaplan, J.H.: Anion diffusion across artificial lipid membranes: The effects of lysozyme on anion diffusion from phospholipid liposomes. Biochim. biophys. Acta *290:* 339–347 (1972).

18   King, R.J. and Clements, J.A.: Surface active materials from dog lung. II. Composition and physiological correlations. Am. J. Physiol. *223:* 715 (1972).

19   Kirkland, V.L.; Reisen, W.H., Gross, A.M., and O'Neill, H.J.: Surface activity of various fractions of rabbit lung lavage fluid. Compar. Biochem. Physiol. *47:* 1077 (1974).

20   Klaus, M.H.; Clements, J.A., and Havel, H.J.: Composition of surface active material isolated from beef lung. Proc. natn. Acad. Sci. USA *47:* 1858 (1961).

21   McDougall, J. and Smith, J.F.: The development of the human type II pneumocyte. J. Path. Bact *115:* 245–251 (1975).

22   Michaelson, D.M.; Horwitz, A.F., and Klein, M.P.: Transbilayer asymmetry and surface homogeneity of mixed phospholipids in cosonicated vesicles. Biochemistry, N.Y. *12:* 2637–2645 (1973).

23 Munden, J.W. and Swarbrick, J.: Time-dependent surface behavior of dipalmitoyllec-
    ithin and lung alveolar surfactant monolayers. Biochim. biophys. Acta *219:* 344–350
    (1973).
24 Obladen, M.; Merritt, T.A., and Gluck, L.: Alterations in tracheal phospholipid compo-
    sition during RDS. Pediat. Res. *10:* 465 (1976).
25 Obladen, M.: Factors influencing surfactant composition in the newborn infant. Eur.
    J. Pediat. *128:* 129–143 (1978).
26 Obladen, M.; Brendlein, F., and Krempien, B.: Surfactant substitution. Eur. J. Pediat.
    131: 219–228 (1979).
27 Papahadjopoulos, D.: Phospholipid model membranes. Biochim. biophys. Acta *135:*
    624–638 (1967).
28 Pattus, F.; Desnuelle, P., and Verger, R.: Spreading of liposomes at the air/water inter-
    face. Biochim. biophys. Acta *507:* 62–70 (1978).
29 Pfleger, R.C.; Henderson, R.F., and Waide, J.: Phosphatidylglycerol – a major compo-
    nent of pulmonary surfactant. Chem. Phys. Lipids *9:* 51 (1972).
30 Pfleger, R.C. and Thomas, H.G.: Beagle dog pulmonary surfactant lipids. Lipid compo-
    sition of pulmonary tissue, exfoliated lining cells, and surfactant. Archs int. Med. *127:*
    863 (1971).
31 Reifenrath, R. and Zimmermann, I.: Dynamic surface tension properties of mixed
    lecithin-cholesterol films related to the respiratory mechanics. Respiration *33:* 303–314
    (1976).
32 Rooney, S.A.; Canavan, P.M., and Motoyama, E.K.: The identification of phosphat-
    idylglycerol in the rat, rabbit, monkey and human lung. Biochim. biophys. Acta *360:* 56
    (1974).
33 Rooney, S.A.; Page-Roberts, B.A., and Motoyama, E.K.: Role of lamellar inclusions in
    surfactant production: studies on phospholipid composition and biosynthesis in rat and
    rabbit lung subcellular fractions. J. Lipid Res. *16:* 418–425 (1975).
34 Stratton, C.J.: The ultrastructure of multilamellar bodies and surfactant in the human
    lung. Cell Tiss. Res. *193:* 219–229 (1978).
35 Tyrell, D.A.; Heath, T.D.; Colley, C.M., and Ryman, B.E.: New aspects of liposomes.
    Biochim. biophys. Acta *457:* 259–302 (1976).
36 Van Deenen, L.L.M.: Phospholipide: Beziehung zwischen ihrer chemischen Struktur
    und Biomembranen. Naturwissenschaften *59:* 485–491 (1972).

PD Dr. med. M. Obladen, Neonatologische Abteilung, Universitäts-Kinderklinik,
Rümelinstrasse 19–23, D-7400 Tübingen (FRG)

Discussion

Following a question by *Robertson, Obladen* discussed the problem of the transition temperature, which for PG is usually 23 °C as opposed to PC. This why he used a disaturated PG which has a very high transition temperature. He considers that for this reason the spreading effect is more likely due to negative charge. The transition temperature of the mixture is 46 °C, so, theoretically, it should not spread. Replying to a question by *Zänker, Obladen* explained that he used different types of liposomes; a crude unshaken multilamellar structure which was made according to *Bangham,* and coworkers, liposomes made by the above-mentioned group by ultrasound at room temperature and the preparation produced according to *Gregoriades* and his group. The liposomes he used had diameters between 200 and 600 Å. Answering a question by *Enhorning, Obladen* stated that the suspension fluid was normal isotonic saline. *Van Golde,* referring to a paper by *Akino and Okano,* said that saturated PG makes up 30% of the pulmonary surfactant which is much lower than the saturation of PC. *Obladen* pointed out that in his opinion it is not DPPG which modifies the PC function, but rather PG itself. It is not known yet whether it acts by lowering the transition temperature, which is unlikely from Obladen's experiments, or by changing the charge of PC. *Clements* and *Obladen* agreed that the behavior of liposomes would be greatly changed if they were prepared in a different way.

Prog. Resp. Res., vol. 15, pp. 188–193 (Karger, Basel 1981)

Physical Properties of Surfactant under Compression[1]

Colin Morley and Alec Bangham[2]

Department of Paediatrics, University of Cambridge, and Biophysics Unit, ARC, Babraham, Cambridge

In this paper, we would like to show that when pulmonary surfactant is repeatedly and gently compressed and expanded, in a manner that might imitate the breathing movements of the lungs, its chemical composition is refined. When this refined monolayer has been formed and is then further compressed it changes its physical state from a liquid to a solid [1].

We would like to postulate that this refinement of surfactant takes place on the alveolar surface and that the alveoli are held open during expiration by a virtually solid monolayer of surfactant phospholipids.

This idea is contrary to the present theories on the action of surfactant. The conventional theory is that pulmonary surfactant forms extremely low surface tensions (10–0 mN m^{-1}) under compression and this counteracts the forces leading to alveolar collapse. This theory was initiated by *Pattle* [2] who observed that surfactant bubbles in water were very small and therefore should have collapsed but were remarkably stable. He calculated correctly from the size of the bubbles that the surface tension of the surfactant layer must be extremely low. However, he did not actually measure surface tension nor, like any of his contemporaries did he consider that the surface film may change in state from liquid to solid. This change in state would have altered all his mathematical calculations. The solidification of surfactant was observed by *Pattle* [3], but not interpreted as such when he blew surfactant bubbles in deaerated water. The pressure

[1] supported by a grant form MRC, Birthright, and the National Fund for Research into Crippling Diseases.
[2] *C.M.* acknowledges the help and encouragement of Prof. *J.A. Davis.*

across the bubble steadily increases as air is absorbed into the surrounding water. This will compress the surfactant film further. Suddenly the bubbles deform (or click). We think that this is due to the compressed solid surfactant film suddenly buckling under the extreme pressure.

Very low surface tensions have been postulated to explain why small alveoli do not collapse into the big ones. It is suggested that in the small alveoli the surfactant is more compressed and therefore has a lower surface tension and withstands the compressive forces thereby stabilising the alveoli [4]. However, if the surfactant layer remains a liquid the molecules in the areas of high compression will quickly move to the areas of lower compression. This will soon equilibrate the surface tensions throughout the alveoli and the lungs will collapse. However, if the surfactant mono-layer solidified on compression it would not matter whether the alveoli were small or large they would be splinted open.

Many workers have used a modified Wilhelmy balance to show that the surface tension of a surfactant monolayer, under compression, falls steadily as the surface is compressed. When they have compressed the surface area by about 80% the measured surface tension is shown to be very low [5]. This method of measuring surface tensions depends upon the surface film forming a meniscus at the dipping plate. It is the pull of the meniscus on the plate which produces the surface tension. If under compression the surface tension of the surfactant film falls to virtually zero what are the physical properties of that surfactant film? How can a zero surface tension be explained by a liquid? We would like to suggest that the surface tension only appears to fall to zero because the film becomes solid and no longer behaves as a liquid. There is no meniscus, therefore no pull on the plate and a spurious reading of a low surface tension is produced.

What evidence have we for this idea of solidification? Let us consider the molecule orientation of a surfactant monolayer at the air liquid inter-face. The main constituents are phospholipids. These have a hydrophilic polar head group and very hydrophobic fatty acid chains. At the surface they will self-assemble so the fatty acid chains are in the air and the polar groups in the water [6] (fig. 1). The surface tension produced by these mole-cules at the aqueous interface will depend on the number of molecules present. If there is an excess of molecules present so that the water is completely covered with phospholipids the surface tension will be reduced from that of water 72 mN m^{-1} to that of hydrocarbon or fatty acid chains $\simeq 25$ mN m^{-1}. This is the equilibrium state for surfactant phospholipids. In this state molecules are tightly packed together. No more molecules can

Fig. 1. A complete phospholipid monolayer. A diagrammatic representation of the orientation and self-assembly of phospholipid molecules at an air/water interface.

Fig. 2. A diagrammatic representation of an incomplete monolayer of mixed surfactant lipids at the air/water interface. With insufficient molecules on the surface, the surface tension rises because water appears.

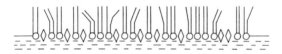

Fig. 3. A diagrammatic representation of a closely packed equilibrium monolayer of surfactant lipids. The water is completely covered so only the surface tension of lipid is measured: approximately 25 mN m^{-1}.

spread onto the surface. The surface tension will not spontaneously fall below this level. Any intermediate surface tension will be due to insufficient molecules being on the surface to cover up the water (fig. 2). The surface film will be incomplete.

When a surfactant film is compressed on a trough the molecules pack close together. If the surface film has been incomplete the surface tension will fall steadily with compression until the surface is covered with a close packed monolayer at a surface tension of approximately 25 mN m^{-1} (fig. 3). Further compression then results in the reduction of area with very little change in surface tension. This may be due to two things. First, the intramolecular structure of the phospholipids may alter slightly to allow tighter packing. Secondly, under the increased pressure the asymmetrical molecules are squeezed out of the surface refining the molecular species of the surface to the more stable molecules (fig. 4). The stable phospholipids are those with saturated fatty acid chains of which dipalmitoyllecithin is the most common in surfactant.

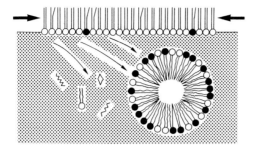

Fig. 4. When a monolayer of surfactant is compressed the surface is refined. Only the stable molecules remain on the surface. The unstable molecules are squeezed out.

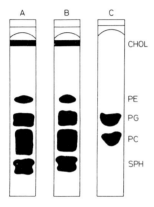

Fig. 5. Thin-layer chromotography. A = Sheep surfactant. B = Sheep surfactant spread as a monolayer – there is no difference in composition with spreading. C = sheep surfactant after repeated compression and expansion at 15% of surface area in the presence of excess surfactant – only lecithin and phosphatidylglycerol remain on the surface. CHOL = Cholesterol; PE = phosphatidylethanolamine; PG = phosphatidylglycerol; PC = phosphatidylcholine; SPH = sphingomyelin.

On expansion of the surface film, in the presence of excess and rapidly available surfactant molecules, as from a dry particle or fresh lamellar bodies, more heterogenous molecules spread onto the surface until the equilibrium surface tension is reached. With repeated compression, expansion and donation of new molecules the surface becomes covered with stable molecules.

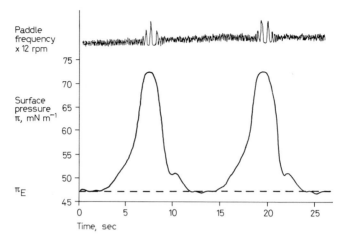

Fig. 6. The surface pressure of DPL/uPG increases when compressed by 40% of the area. This causes the paddle of the rotating viscometer to stop turning.

We have shown, by analysing surfactant from a surface film before compression and after repeated compression and expansion through about 15–20% of the surface area, that the analysis of the surface film has changed. It has become refined to lecithins and phosphatidylglycerols (fig. 5). Further and more careful experiments may show that the surface refines to pure dipalmitoyllecithin – physically an extremely stable molecule that is capable of resisting considerable forces without buckling.

If this refined surfactant is now compressed by about 40% of its area. The surface solidifies. This can be shown in several ways. If the surface film is closely observed it takes on a dull sheen as the surface solidifies. It can also be demonstrated by the fact that a fine dusting of talc on the surface will no longer move easily when the film is compressed. We have shown this using a light rotating paddle as a surface viscometer [1]. When the film is compressed by more than 40%, the surface pressure rises, the viscocity increases and the paddle stops rotating (fig. 6). This indicates a considerable change in the physical state of the monolayer and demonstrates that even under moderate compression the surfactant monolayer will solidify.

In the lung the surfactant monolayer is repeatedly compressed and expanded. It is therefore possible that *in vivo* surfactant is refined and solidifies as the alveoli begin to collapse. This solid layer of molecules may then prevent atelectasis.

Summary

Under repeated compression pulmonary surfactant is refined to lecithin and phosphatidylglycerol. When compressed by more than 40% the monolayer solidifies. This may be the antiatelectasis mechanism of surfactant.

References

1   Bangham, A.; Morley, C., and Phillips, M: The physical properties of an effective lung surfactant. Biochim. biophys. Acta *573:* 552–556 (1979).
2   Pattle, R.: Properties, function and origin of the alveolar lining. Proc. R. Soc. B. *148:* 217–240 (1958).
3   Pattle, R.: Surface lining of lung alveoli. Physiol. Rev. *45:* 48–79 (1965).
4   Reynolds, E., and Strang, L.: Alveolar surface properties of the lung in the newborn. Br. med. Bull *22:* 79–83 (1966).
5   King, R. and Clements, J.: Surface active materials from dog lung. I. Method of isolation. Am. J. Physiol *223:* 707–714 (1972).
6   Bangham, A.: Physical structure and behaviour of lipids and lipid enzymes. Advances in lipid research, pp. 65–104 (Academic Press, New York 1963).

C. Morley, MD, Department of Paediatrics, University of Cambridge, Cambridge (England)

Discussion

Clements wondered what happened to the substances, since *Morley* had only recovered DPPC. *Morley* believed that sphingomyelin and phosphatidylethanolamine may go to the subphase. Answering a question by *Tierney, Morley* pointed out that the refinement of film might have been overlooked for many years because the film compression was mostly done to 80% and this refinement would only happen if the film is compressed less than 50%. The compression on the film, going back to work by *Gaines,* is high. 1 dyne centimeter acting on a film 28 Å high was equivalent to 4,9 atmospheres and if we apply an excess pressure of 25 dynes that means that these films have to sustain, at least for a period of time, something in excess of 100 atmospheres which is really quite remarkable.

Clements said that even at total lung capacity the surface tension was not higher than 30 dynes per centimeter, so it seems that the film in the lungs would be formed pretty close to the equilibrium film density. The necessary area change, therefore, would be small in order to compress the film to very low surface tension.

Morley pointed out that surfactant became solid if compressed by 50%. In a normally breathing lung the changes would be 10–15%, from this point of view it may be that the solidification only occurs in very rare situations, for instance by opening the chest, but if there is a refinement of surfactant to a pure DPPC, with a transition temperature of 41 °C, this refined film at body temperature will be a solid substance and no longer a liquid.

Prog. Resp. Res., vol. 15, pp. 194–206 (Karger, Basel 1981)

Effect of Ventilation on the Surface Properties of the Lung

W. Mitzner and S. Permutt

Divisions of Environmental Physiology and Medicine, The Johns Hopkins University School of Hygiene and Public Health, Baltimore, Md.

Before beginning this discussion we would like to acknowledge the efforts of Drs. *J.W.C. Johnson* and *E.E. Faridy,* each of whom played a major role in the generation and analysis of the data to be presented.

In order to talk about lung stability we first have to carefully define a measure of this stability. Ideally, one would like to be able to examine a single alveolus and measure its size at various transpulmonary pressures. However, this is not practical and we therefore must infer the behavior of a single alveolus from the behavior of many alveoli. Were all alveoli identical this would present no problem. But in fact, alveoli are not identical, and even at the same transpulmonary pressure there is a wide variation in alveolar size. Thus, to quantify alveolar stability by looking at the whole lung, one requires certain assumptions. The basic assumption used by nearly all investigators in this area is that at some defined maximum lung volume, if one observes a change in this maximal volume, then this change has occurred only by a change in this number of alveoli inflated. That is, we assume that when the lung is at maximal volume (V_{max}), each alveolus is also at maximal volume, and that each alveolus is no longer distensible. This assumption implies that a decrease in V_{max} can only occur as a result of some atelectasis in the lung. With this assumption we can examine the deflation pressure-volume behavior of only those alveoli which are inflated. This is done by normalizing the pressure-volume curve by calling V_{max} the 100% volume, and expressing all volumes as %V_{max}.

To analyze the normalized curves we need to make the additional assumption that the pressure-volume behavior of each alveolus is independent of the behavior of other alveoli. Thus, even though there may be some atelectasis, we assume that this has no effect either on the mean

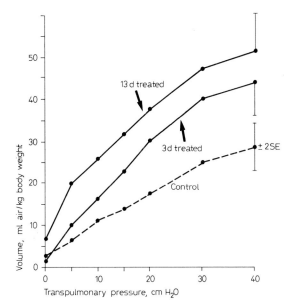

Fig. 1. Mean deflation limbs of air pressure-volume curves for control and betamethasone-treated rhesus monkey fetuses all delivered at 133 days gestational age. Volume is expressed as milliliter air per kg body weight. Lung weight/body weight ratio is unaffected by betamethasone.

alveolar size or the pressure-volume behavior of the open units. With these assumptions the deflation limb of the normalized pressure-volume curve represents the average behavior of individual alveoli, and any change in the percent volume at a specific transpulmonary pressure is interpreted as a change in the properties of the surface lining layer.

To illustrate this method as well as the basic problem in interpretation, which we have only recently come deal with, I would like to digress briefly and show some results where we have attempted to look at the mechanism of a similar but opposite situation, that is the augmentation of fetal lung stability by glucocorticoids. For this purpose I will only show summarized results. More specific details can be found elsewhere [6, 7]. We studied lungs from rhesus monkey fetuses at an average gestational age of 80% pregnancy term. Betamethasone was maternally administered 3 or 13 days prior to cesarian delivery. Shortly after fetal death the lung pressure-volume curves were done with both air and saline inflation. Also, in another series, the lungs were fixed at maximal volume, and morphometric

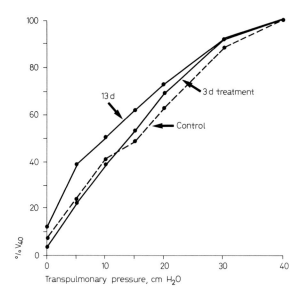

Fig. 2. Deflation limbs from figure 1 plotted as a percent of the volume at 40 cm H₂O for each curve.

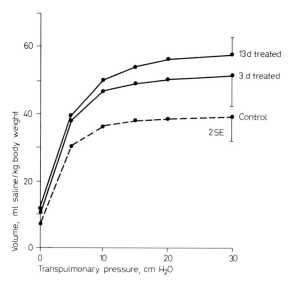

Fig. 3. Mean deflation limbs from the saline pressure-volume curves, from the same animals as in figure 1. Volume is expressed as milliliter saline per kg body weight.

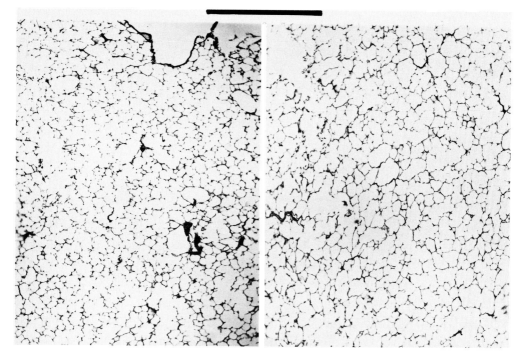

Fig. 4. Low power view of 7 μm-sections from control (left) and betamethasone-treated (right) 130-day rhesus fetal lungs. Lungs were fixed at 30 cm H_2O transpulmonary pressure. Solid bar at top of figure = 1 mm.

analysis of alveolar size was done. Figure 1 shows the averaged pressure-volume curves with air. As can be seen, one of the major changes of betamethasone is on the V_{max}, defined as the volume at 40 cm H_2O. To assay for surfactant we plot the deflation limbs as a %V_{max}. This is shown in figure 2. As can be seen, there is some evidence for increased deflation stability following steroid treatment. That is, the volumes remaining on deflation are greater. From this and other work we might conclude that there is increased surfactant present. However, is this increased stability really independent of the change in V_{max}? Figure 3 shows the results of saline inflation. As can be seen from figure 3 much of the increase in V_{max} following steroids exists even when the lungs are inflated with saline. Thus, the major part of the increased V_{max} has little to do with surfactant, and must be related to tissue structural changes. Even so, we still might be able

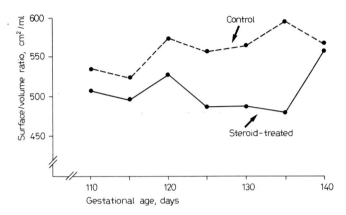

Fig. 5. Morphometrically determined surface/volume ratio in cm²/ml as a function of gestational age for control and betamethasone-treated fetal rhesus monkeys. There is no significant trend with gestational age, but the mean of all the steroid-treated animals is significantly less (p<0.05) than the mean of the control animals.

to assay for functional surfactant changes if we could show that the alveoli are the same size at maximal lung volume in control and treated groups. That is, if this increase in V_{max} occurred by recruitment and not by stretching, we still might be all right. This analysis was not done for the lungs shown, but morphometric analysis was done in another series of monkey lungs fixed at maximal lung volume. In this series, pairs of fetal animals were examined at gestational ages from 110 to 140 days. Figure 4 shows the results from the 130-day animals. The overall visual perspective of this figure clearly indicates larger airspace size in the glucocorticoid-treated animal. Figure 5 shows the results of the morphometric analysis of surface area density for all animals. As can be seen, the treated animals show a decreased surface area/volume ratio, consistent with larger alveoli. We also measured the alveolar numerical density and showed an average decrease of 21% in alveolar number per milliliter following betamethasone. These findings essentially invalidate our rigorous use of the normalized pressure-volume curves to assay for surfactant in this case. More importantly, it also emphasizes the need for caution in interpreting the results of volume-normalized lungs.

    In the experiments to be described no attempt was made to determine alveolar size at maximal volume following the different ventilation regimes. Thus, when analyzing for surfactant by examining the normalized pressure-volume curves, we are forced to make the assumption that

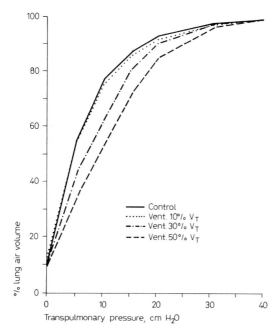

Fig. 6. Volume-normalized deflation limbs showing the effect of duration of ventilation on the deflation stability. Tidal volume = 30% V_{max}.

alveolar size at V_{max} is not affected by prior ventilation. This assumption could be tested by fixing the lungs at maximal volume, as was done with the monkey lungs, and the ultimate interpretation of these data will depend on the results of such experiments.

The basic series of ventilation experiments were done on isolated dog lobes. The lobes were ventilated at a fixed rate of 12 breaths/min, and tidal volume, duration of ventilation, and the level of positive end-expiratory pressure were varied. The lungs were not perfused with blood, but as I will show, the metabolic activity of the epithelial cells is apparently not impaired by this lack of blood perfusion. Indeed, as *Riley* [9] pointed out some years ago, the lung is the one organ that has a higher PO_2 in the absence of circulation than in the presence of circulation.

Figure 6 shows the effect on the normalized deflation limb of the quasi-static pressure-volume curve of varying the length of time of ventilation. The curve-labelled control was done prior to any mechanical ventilation and the effect of 2, 3, and 6 h of ventilation (tidal volume = 30% V_{max}). As

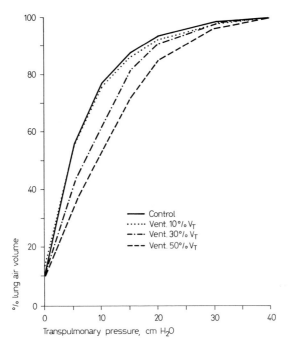

Fig. 7. Volume-normalized deflation limbs showing the effect of tidal volume, V_T, on the deflation stability. Duration of ventilation = 3 h.

can be seen, the normalized deflation curves are depressed the longer the ventilation time. In some subsequent analysis, I will refer to a single point representative of each curve. This is the percent volume remaining at 10 cm H_2O (notation: $\%V_{10}$). If the $\%V_{10}$ is smaller, then the lung has less stability, and we interpret this to mean there is less surfactant. Figure 6 shows that at least under our experimental conditions ventilation decreased lung stability and the extent of this decrease in a function of how long the ventilation lasts.

Figure 7 shows the effect of changing the tidal volume. The axes are same as figure 6. Lungs were ventilated for 3 h at 10, 30, 50% V_{max}. Figure 7 shows quite clearly that the greater the tidal volume the greater the fall in lung stability. Although the deflation stability is decreased immediately following ventilation, this decreased stability can be prevented by applying positive end-expiratory pressure to the lung during the ventilation period. Figure 8 shows the $\%V_{10}$ taken from the pressure-volume curves immedi-

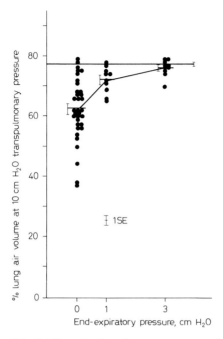

Fig. 8. Effect of end-expiratory pressure on the deflation stability parameter, $\%V_{10}$.

ately following ventilation as a function of the end-expiratory pressure used during the ventilation period. In this case the ventilation time was 3 h and the tidal volume was 30% V_{max}. Figure 8 shows that the application of end-expiratory pressure can prevent the decrease in lung stability caused by the ventilation. That is, at increasing end-expiratory pressure the $\%V_{10}$ approaches the control value.

Although ventilation without end-expiratory pressure results in a decreased deflation stability, this decreased stability can be reversed by stopping the ventilation and leaving the lungs inflated at a constant distending pressures. Figure 9 shows how the deflation pressure-volume curves are altered by a 3-hour period of steady inflation following ventilation. As can be seen, the deflation curves of control and ventilation followed by static inflation are essentially identical. This is in marked contrast to the effects of ventilation alone which shows a significantly depressed curve. Thus, static inflation apparently allows some surface

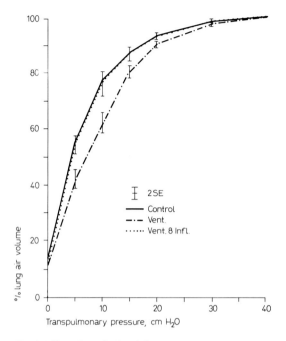

Fig. 9. Effect of ventilation followed by 3 h steady inflation at 30 cm H$_2$O on the normalized deflation limbs.

recovery to occur, and the stability of the lung is increased back to the control level.

The metabolic nature of the decreased stability caused by ventilation and the recovery caused by static inflation were investigated by performing similar experiments but varying the fraction of O$_2$ in the ventilating gas. Figure 10 shows the %V$_{10}$ of the quasi-static pressure-volume curve following ventilation alone or ventilation plus static inflation as a function of the percent oxygen in the ventilating gas. This graph shows that the extent of recovery of stability is a direct function of the fractional concentration of O$_2$ in the ventilating gas. That is, at lower PO$_2$ values the recovery (which is the difference between the lines of ventilation alone and ventilation + inflation) is markedly impaired, whereas at high PO$_2$, the recovery is enhanced. Indeed, at 100% O$_2$ even the fall in %V$_{10}$ with ventilation alone is not as great as with low oxygen. This oxygen dependence of the recovery of stability is consistent with an active metabolic process. The metabolic nature was also shown by additional studies where the

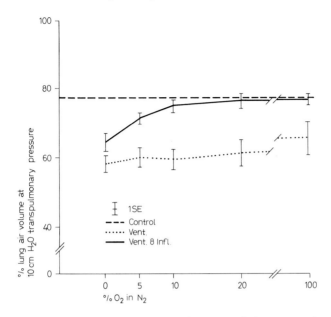

Fig. 10. Effect of varying the %O_2 in the ventilating gas on the deflation stability parameter, %V_{10}, for lungs ventilated for 3 h only, and for lungs ventilated for 3 h followed by a 3-hour static inflation with room air.

temperature was changed during the recovery period. These results showed markedly impaired recovery of stability at low temperatures [1].

Several other investigators have presented results which support those presented here. *McClenahan and Urtnowski* [5] found a decrease in the %V_{15} following ventilation in both rat and dog lungs. They also suggest an active metabolic recovery process since the recovery was inhibited by KCN and low temperature. Furthermore, they also investigated the effect of varying the rate of ventilation, something which we did not study. They found that the decrease in stability was enhanced at higher rates. In fact, they claim that the ventilation rate is a more important parameter than the length of time of ventilation.

Similar results were also found by *Forrest* [2] in intact guinea pigs mechanically ventilated with large tidal volumes (up to 75% VC). In a morphometric study *Forrest* [3] also suggested that hyperventilation results in a decreased alveolar volume with no change in alveolar duct volume. However, this inhomogeneous alteration in alveolar size found by

Forrest [3] was not supported by the study of *Young et al.* [12]. These investigators suggested that the decreased lung volume at low transpulmonary pressure was not caused by decreased alveolar size but rather by small areas of atelectasis. *Williams et al.* [10] also showed some atelectasis following 1 h of mechanical ventilation in intact animals.

In intact dogs, *Greenfield et al.* [4] showed that hyperventilation results in increased surface tension of lung extract. In intact open-chest cats, *Wyszogrodski et al.* [11] showed that hyperventilation decreased the $\%V_{15}$ as measured on the subsequent quasi-static pressure-volume curve. As was found here, this decreased stability was prevented by the application of 2.5 cm H_2O end-expiratory pressure. The decrease in stability was augmented by increasing the tidal volume and the length of time of ventilation. Also, as was found by *McClenahan, and Urtnowski* [5] they found that increasing the respiratory rate markedly decreased the stability as well.

In considering an intact living lung, we know that there is a continual synthesis and degradation of surfactant. One important aspect of this synthesis pointed out by *Wyszogrodski et al.* [11] and later amplified by *Oyarzun and Clements* [8] relates to the effect of ventilation on surfactant production on release. These investigations have shown that ventilation itself acts to activate increased levels of alveolar surfactant. Such being the case, this makes the results presented here even more striking, since were there no activation of surfactant occurring, the loss of stability we observed with ventilation would have been even greater. The studies reported here and the similar results of others show which conditions accelerate the degradation process. With such concitions the equilibrium between production and degradation which exists in a normal healthy lung is no longer present.

One final word is warranted relating the significance of the results presented to the focus of this symposium, that is the clinical relevance. Since mechanical ventilation is widely used in intensive care, and since decreased lung stability in this situation is especially bad, we might summarize our results by considering those factors affecting optimal mechanical ventilation to minimize the loss of lung stability due to ventilation. There are at least three factors to consider:

(1) The level of positive end-expiratory pressure. In this study with normal lungs we found that a low level of end-expiratory pressure was sufficient to preserve lung stability. However, whether this is true in diseased lungs or whether a higher level is required remains to be shown.

(2) The magnitude of total ventilation. Avoid hyperventilation, whether by increased tidal volume or increased rate. (3) The level of alveolar oxygen. Hypoxia accelerates the loss of stability. However, in a diseased lung with very uneven $\dot{V}/\dot{Q}$ distribution, the minimum level of inspiratory O_2 required needs to be determined in each case.

Such optimization of ventilation may help to avoid or at least minimize the occurrence of alveolar atelectasis and resultant hypoxemia often associated with mechanical ventilation during and after surgical anesthesia.

Summary

Experimental conditions are described where mechanical ventilation has a deleterious effect on lung stability. These conditions are large tidal volumes, high ventilation rates, low oxygen, low temperature, and low end-expiratory pressures. To interpret this loss of stability as being the result of surfactant degradation requires certain assumptions regarding the analysis of the volume normalized deflation limb of the pressure-volume curve. The limitations of these assumptions are discussed with reference to another series of experiments dealing with the stability of fetal monkey lungs following glucocorticoid treatment. In this case the assumptions are not valid, and the functional properties of surfactant are masked by significant changes in tissue structure and alveolar size.

References

1   Faridy, E.E.; Permutt, S., and Riley, R.L.: Effect of ventilation on surface forces in excised dog lung. J. appl. Physiol. *21:* 1453–1462 (1966).

2   Forrest, J.B.: The effect of hyperventilation on pulmonary surface activity. Br. J. Anaesth. *44:* 313–320 (1972).

3   Forrest, J.B. The effect of hyperventilation on the size and shape of alveoli. Br. J. Anaesth. *42:* 810–817 (1970).

4   Greenfield, L.J.; Ebert, P.A., and Benson, D.W.: Effect of positive pressure ventilation on surface tension properties of lung extracts. Anaesthesia *25:* 312–316 (1964).

5   McClenahan, J.B. and Urtnowski, A.: Effect of ventilation on surfactant, and its turnover rate. J. appl. Physiol. *23:* 215–220 (1967).

6   Mitzner, W.; Johnson, J.W.C.; Hutchins, G.; Beck, J.; Scott, R.; Palmer, A., and London, W.: Morphometric analysis of fetal rhesus monkey lungs following treatment with betamethasone. Fed. Proc. *38:* 964 (1979).

7   Mitzner, W.; Johnson, J.W.C.; Scott, R.; London, W.T., and Palmer, A.E.: Effect of betamethasone on pressure-volume relationship of fetal rhesus monkey lung. J. appl. Physiol. *47:* 377–382 (1979).

8   Oyarzun, M.J. and Clements, J.A.: Control of ventilation, adrenergic mediators, and
     prostaglandins in the rabbit. Am. Rev. resp. Dis. *117:* 879–891 (1978).
9   Riley, R.L.: Apical localization of pulmonary tuberculosis. Bull. Johns Hopkins Hosp.
     106: 232–239 (1960).
10   Williams, J.V.; Tierney, D.F., and Clements: Surface forces in the lung, atelectasis, and
     transpulmonary pressure. J. appl. Physiol. *21:* 819–827 (1966).
11   Wyszogrodski, I.; Kyei-Aboagye, K.; Taeusch, H.W., and Avery, M.E.: Surfactant inac-
     tivation by hyperventilation: conservation by end-expiratory pressure. J. appl. Physiol.
     38: 461–466 (1975).
12   Young, S.L.; Thierny, D.F., and Clements, J.A.: Mechanism of compliance change in
     excised rat lungs at low transpulmonary pressure. J. appl. Physiol. *29:* 780–785 (1970).

W. Mitzner, PhD, Divisions of Environmental Physiology and Medicine, The Johns
Hopkins University School of Hygiene and Public Health, Baltimore, MD 21205 (USA)

Discussion

Answering a question by *von Wichert, Mitzner* pointed out that not only a high tidal
and low rate ventilation but also high rate at low tidal ventilation would have a bad effect upon
the surfactant, so it seems that the problem is hyperventilation. *Tierney* reported investigations together with *Webb,* showing that alveolar pulmonary edema induced in rats by ventilating the animal with high pressure (40–45 cm H_2O) could be prevented by introducing end-
expiratory pressure. These animals developed interstitral pulmonary edema but did not die
and did not develop interalveolar edema as those animals without end-expiratory pressure.
Tierney thinks that it is not caused by a depletion of surfactant but by a change in surface
tension of the alveoli. *Falke* was careful in applying these results to patient care, because there
is no proof that patients with low tidal ventilation would be better than those with high tidal
ventilation. *Falke* asked *Mitzner* if the low effective compliance using low tidal volumes could
be explained by surface properties. *Mitzner* answered that low tidal volume ventilation would
lead to some degree of atelectasis. *Riefenrath,* based on investigations done with *Reiss* and
Petty, feels that surfactant is more important on the bronchial level than on the alveolar level,
because ventilation with high tidal volume leads to alveolar closure.

Prog. Resp. Res., vol. 15, pp. 207–218 (Karger, Basel 1981)

Surface Properties during Mechanical Ventilation and Their Clinical Estimation

M. Baum, H. Benzer, A. Geyer, N. Mutz and G. Pauser.

Department of Intensive Care Therapy (Head: Prof. *H. Benzer*) of the Clinic for Anaesthesiology and General Intensive Medicine (Director: Prof. *O. Mayrhofer*) and the 2nd Surgical Clinic (Prof. *J. Navratil*) of the University of Vienna, Vienna

From the present point of view it seems that the surfactant system is rather sensitive to mechanical stress. Although the lung surfactant is able to cope with normal spontaneous breathing patterns, mechanical ventilation in some way can disturb the surface tension properties. From animal experiments we have learned that hyperinflation with large tidal volumes and thus alveolar pressures will deteriorate the surface activity [3–5].

Low transpulmonary pressures, on the other hand, will lead to similar effects. Both conditions can be met during mechanical ventilation, as the tidal volume is no longer uniformly distributed over the lungs. To some extent this can be counteracted by a careful setting of the ventilator, which avoids high inflation pressures during inspiration. A selective application of positive end-expiratory pressures (PEEP) may compensate for a reduced functional residual capacity (FRC).

Unfortunately, there is no general rule for what the optimal settings are. They highly depend on the mechanical and gas exchange properties of the lung being ventilated. To adjust a surfactant-compatible ventilation pattern one needs a measure for the surface activity during mechanical ventilation. The most convenient approach is given by pressure/volume (p/V) studies of the lungs. This information can be derived either from static pressure-volume diagrams – which is the traditional way – or dynamically from the pressure-volume properties during mechanical inflation by a ventilator.

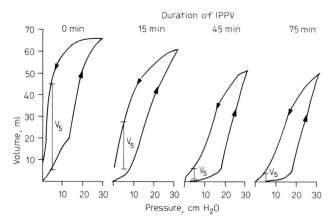

Fig. 1. Initial p/V diagram and p/V diagrams of rabbits after 15, 45 and 75 min. of mechanical ventilation (IPPV).

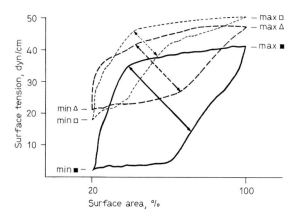

Fig. 2. Lung tissue extracts in the Wilhelmy balance after increasing duration of mechanical ventilation. max = Maximum surface tension; min = minimum surface tension.

In the following we shall focus on the influence of PEEP on the surfactant system and on different methods of evaluation of surface properties during mechanical ventilation.

12 years ago we looked for changes in surface tension during mechanical ventilation of rabbits with open chest. The ventilation was performed by a time-cycled constant-pressure ventilator with inspiratory pressures of

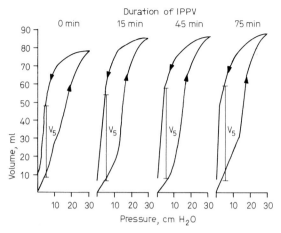

Fig. 3. Sequence of pressure-volume diagram during mechanical ventilation with PEEP of 5 cm H_2O.

10 cm H_2O and an expiratory pressure of zero. Figure 1 shows the initial p/V diagram and p/V diagrams of the animals after 15, 45 and 75 min of this treatment. The deflation limb of every diagram characterizes the surface tension properties. Normal surface activity will cause a high volume to be retained in the lungs even with low alveolar pressures. Among other parameters, V5 – the volume trapped with 5 cm H_2O at the deflation limb – is a simple measure for surface activity. During the experiment a marked decrease of V5 was observed, indicating a successive loss of surfactant. The convex shape of the deflation section of the control p/V diagram becomes more and more concave, which is a quick test for changes of the surfactant system [2].

The same sort of things happened on the Wilhelmy balance, where lung tissue extracts from rabbits killed after increasing duration of mechanical ventilation were exercised (fig. 2). All significant parameters (minimum surface tension, maximum surface tension, and area) indicated a progressive decrease of surface activity. A second group of animals was ventilated in the same way but with an additional PEEP of 5 cm H_2O.

Under these conditions we did not find any deterioration of the surfactant system, demonstrated by the sequence of pressure-volume diagrams of figure 3. We even observed a slight increase of V5 during the ventilation period. The results in the Wilhelmy balance corroborated these findings.

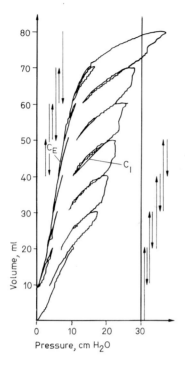

Fig. 4. Modified pressure-volume diagram: pneumoloop representation. $C_Q = C_E/C_I = 3.0$.

As a result of these early studies we assumed that the degree of inflation during the expiratory period mainly influences the stability of the surfactant system. Under open-chest conditions the lung empties during expiration to an unphysiologic low volume which causes an abnormal high degree of compression of the lining layer. A proper level of PEEP restores the normal FRC and thus avoids this additional mechanical stress.

To see wether there was a direct influence of the FRC on the mechanical properties of a normal respiratory loop we started to modify the traditional p/V diagram. Normal p/V diagrams fail to reflect the conditions during artificial ventilation because, in these cases, total expansion of compression of the alveolar surface does not occur. The so-called pneumoloop (fig. 4), which is a modified p/V diagram, gives much more information about these conditions. It consists of a sequential arrangement of respiratory loops that are statically recorded during stepwise changes of the

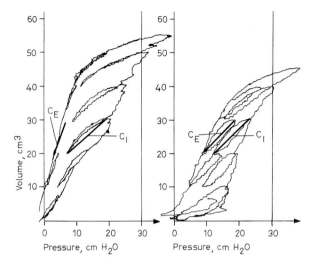

Fig. 5. Pneumoloop diagram in the case of normal (left) and abnormal (right) surface activity.

inspiratory and expiratory FRC. The slope of the individual respiratory loops corresponds to their compliance [1].

As soon as one looks to either the inflation or the deflation limb of this diagram, the individual loops do not differ to much. However, two respiratory loops recorded at the same FRC but during the inflation phase (C_I) and the deflation phase (C_E), respectively, show a different compliance, at which a loop at the deflation limb has a 3 times better compliance. This means that the mechanical properties for a respiratory cycle in the range of normal tidal volumes really depend on the volume history of the lungs. The only system that may change the elastic behaviour in such a way is the surfactant system.

To prove this hypothesis we have compared the pneumoloop diagram of rabbits before and after a lung washing with oleinic acid, which inhibits the surfactant function (fig. 5). After this manœuvre the difference in the slopes of corresponding loops has disappeared; the compliance quotient (C_Q) equals one. Similar results could be obtained from human newborns with and without respiratory distress syndrome. We do not want to go into the theoretical background of this phenomenon, but we would like to demonstrate some clinical applications of this principle.

One can easily convert the pneumoloop diagram into an easy-to-

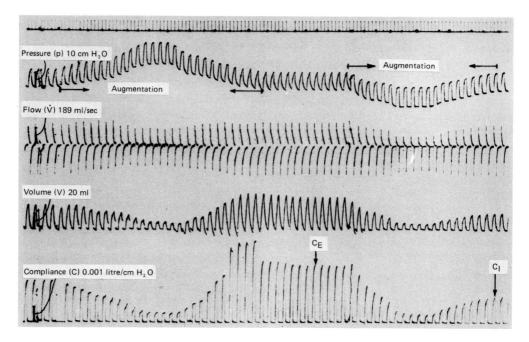

Fig. 6. Pneumotachographic recording during artificial ventilation. Compliance quotient in the case of undisturbed surface activity.

handle bedside evaluation method for surface activity in the mechanically ventilated newborn (fig. 6).

During a period of controlled ventilation, pressure, flow and volume are measured by means of a pneumotachography. An additional hardware calculates the total compliance for each breath. The height of the spikes of the tracing at the bottom of figure 6 corresponds to the magnitude of compliance for each particular breath. In a first manœuvre the PEEP is elevated to 10 cm H_2O and then brought back to its previous level of zero. This, in terms of the pneumoloop diagram, means that the lungs are firstly inflated by consecutive loops with increasing FRC and then deflated again in the same way but with decreasing FRC. The compliance derived after this manœuvre, therefore, corresponds to C_E of the pneumoloop. After that, the end-expiratory pressure is reduced to –10 cm H_2O and again restored to its previous level of zero. The compliance at the end of this period corresponds to C_I of the pneumoloop because it was recorded after an increase of the FRC. In accordance to the pneumoloop we found a

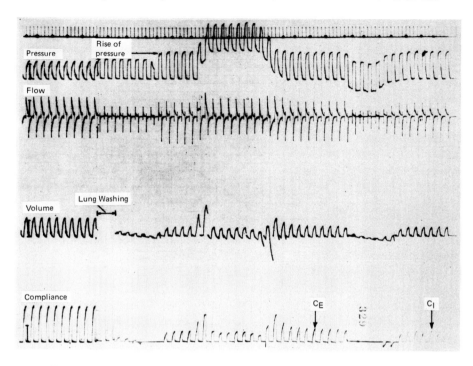

Fig. 7. Pneumotachographic recording during artificial ventilation after a lung washing with oleinic acid.

similar compliance quotient C_E/C_I of 2.7. In contrast, the same sequence of respiratory manœuvres after a lung washing with oleinic acid did not lead to changes in compliance (fig. 7); the quotient C_Q was calculated to be 1.2, which again fits to the values obtained from the pneumoloop diagram. Although this method turned out to be very sensitive in rabbits as well as in human newborns, we could not get similar results in adults.

The different morphometric conditions in the lungs of adults, particularly their larger alveolar diameters, must change the implications of surfactant on the mechanical properties of these lungs. A pneumoloop diagram of a 34-year-old patient without any lung disturbances did not show the characteristic difference in the slopes of the corresponding loops (fig. 8).

As a matter of fact, there is hardly any hysteresis of the p/V properties which in rabbits is said to be due to the surfactant system. Therefore, one

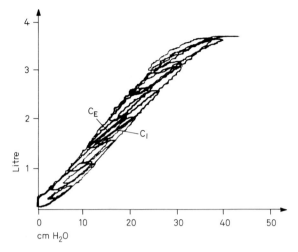

Fig. 8. Pneumoloop diagram of a 34-year-old patient without any lung disturbances.
$C_Q = 1.2$.

must be very careful in extending such experimental results to the human lung physiology.

We presently are studying a modified pneumotachographic method which hopefully will allow us to get some information on the surfactant system of adult patients (fig. 9).

We increase PEEP over 10 min, until we have reached a level of 20 cm H_2O, and return to zero during the following 10 min. All other ventilatory parameters, particularly the tidal volume (V_T), are kept constant during this period. A specially developed electronic hardware samples the actual PEEP level and the corresponding pressure amplitude p. Both signals are fed into an X-Y-plotter, the X-axis recording the PEEP and the Y-axis recording the difference between the end-inspiratory and the end-expiratory pressure. As long as V_T is kept constant, this pressure difference is proportional to the lung elastance.

The original tracing of a dog's lung before and after a lung washing with oleinic acid is shown in figure 10. The inflation sections in both cases are quite different. The elastance of the lung with disturbed surface properties is increased. From our experience with p/V diagrams we do not rely on the inflation limb as a measure for surface activity. However, the interpretation of the deflation limb in this diagram is not trivial. For reason of better understanding, a conversion of the deflation section of the elastance-

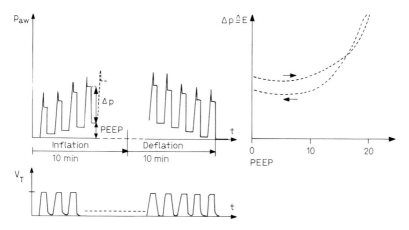

Fig. 9. Modified pneumotachographic method to get information on the surfactant system.

$$E = \frac{\Delta p}{V_T}\,; V_T = \text{Const.} \rightarrow E \,\hat{=}\, \Delta p \,\hat{=}\, \frac{1}{C}\;\cdot$$

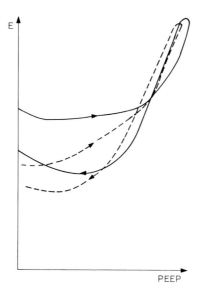

Fig. 10. Modified pneumotachographic method, original tracing of a dog's lung before – – – and after —— a lung washing with oleinic acid.

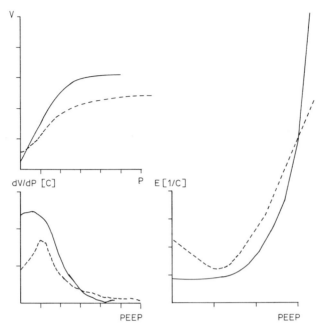

Fig. 11. Conversion of the deflation section of the elastance-PEEP diagram into a compliance-PEEP diagram, performed by a computer program. − − − = Disturbed surface properties.

PEEP (E-PEEP) diagram into a compliance-PEEP (C-PEEP) diagram is performed by a computer program (fig. 11). The dashed line which again represents the disturbed surface properties shows a marked maximum in compliance at a higher level of PEEP, which is not met with normal lung conditions where the maximum compliance is found at PEEP zero. The compliance, on the other hand, may be expressed as the differential quotient dV/dp of the p/V diagram. Looking to the slope of the p/V diagram one can deduce that a convex shape of the deflation section will cause the maximum compliance to be shifted away from the origin in this diagram. With a concave shape, compliance will decrease as soon as transpulmonary pressure is increased. There is good evidence that the presence of a marked compliance maximum, its magnitude and the level of PEEP at which it appears could be a measure of surfactant function during mechanical ventilation in adults.

A recalculation of the E-PEEP diagram by a numerical mathematical method proves the close connection to the p/V diagram (fig. 12). The

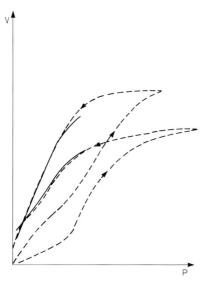

Fig. 12. Recalculation of the elastance-PEEP diagram by a numerical mathematical method.

portion of the deflation limb that can be calculated from the E-PEEP diagram is superimposed to the original p/V diagram of the dog described earlier. In case of normal surface tension as well as under disturbed conditions we found a good correlation between both methods.

References

1   Baum, M. and Benzer, H.: Surface-tension properties and artificial ventilation in the newborn infant. Neonat. pediat. Ventilat. *12:* 61 (1974).
2   Benzer, H.: Respiratorbeatmung und Oberflächenspannung in der Lunge; in Anaesthesiology and Resuscitation, vol. 38 (Springer, Berlin 1969).
3   Farady, E.E.; Permutt, S. and Riley, R.L.: Effect of ventilation on surface forces in excised dogs' lungs. J. appl. Physiol. *21:* 1453 (1966).
4   McClenahan, J.B. and Urtnowski A.: Effect of ventilation on surfactant, and its turnover rate. J. appl. Physiol. *23:* 215 (1967).
5   Webb, H.H. and Tierney D.F.: Experimental pulmonary edema due to intermittent positive pressure ventilation with high inflation pressure, protected by positive end-expiratory pressure. Am. Rev. resp. Dis. *11:* 556 (1974).

Ing. M. Baum, Clinic for Anaesthesia and General Intensive Medicine,
Spitalgasse 23, A-1090 Vienna (Austria)

Discussion

Answering a question by *Dr. von Wichert, Dr. Baum* said that they had only a few figures relating to patients until now. The explanation in species differences in p/V loops and in the cause of life seems to be dependent on the alveolar radius; this special sort of p/V loops is found in animals with low alveolar radius, so he believes that the meaning of the surfactant in adult lung is quite different to what it means in newborn lung; thus it is not necessary to reach in the adult lung such low surface tension as in the lung of the newborn.

Prog. Resp. Res., vol. 15, pp. 219–223 (Karger, Basel 1981)

Alterations in DPL and PL Content of the Lung after Traumatic Shock with Regard to Morphological Changes[1]

A. Lohninger, H. Redl. G. Schlag and A. Nikiforov

Forschungsinstitut für Traumatologie der AUVA, Wien

Attention to the role of the lung in shock is based on experimental and clinical observations of markedly impaired pulmonary function and morphological changes like hemorrhage, edema and atelectasis. More recent studies have demonstrated metabolic alterations of the affected lung tissue involving phospholipid metabolism. In the present study we tried to compare alterations in dipalmitoyllecithin (DPL) and phospholipid content and morphological changes in the lung after drum shock.

Methods

16 rats were divided into 2 groups one control group and the other was shocked by drum. 48 h after this drum shock, half of the lung was ligated at the hilum to allow blood free perfusion. The other half of the lung was installed by a fixative with 12 cm water pressure. The fixed tissue was further processed for transmission and scanning electron microscopy. The blood-free half of the lung was removed immediately after the perfusion and homogenized in 40 vol of chloroform: methanol and extracted under nitrogen overnight. The tissue lipid extracts were washed using the method of *Folch* [2]. The phospholipids were determined according to the method of *Bartlett* [1]. DPL was determined by gas-liquid chromatography with glass-capillary column. This method permits a quantitative DPL determination, without DPL being impurified at all by other PC species [3].

[1] This paper was supported by a grant from the Lorenz-Böhler-Forschungsfonds.

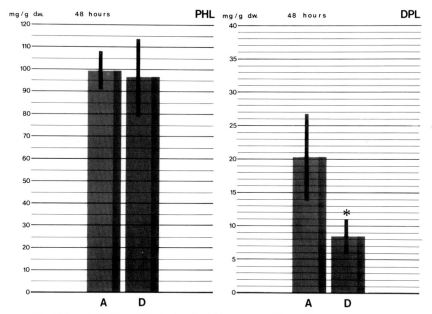

Fig. 1 Phospholipid and dipalmitoyllecithin content of lungs from controls (A) and fed traumatized rats (D).

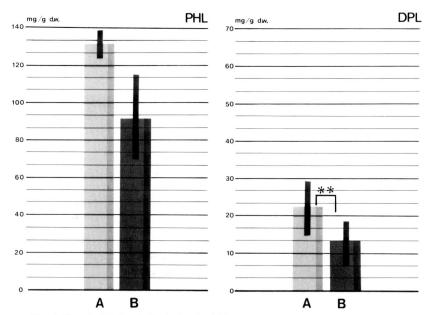

Fig. 2 Phospholipid and dipalmitoyllecithin content of lungs from controls (A) and starved traumatized rats (B).

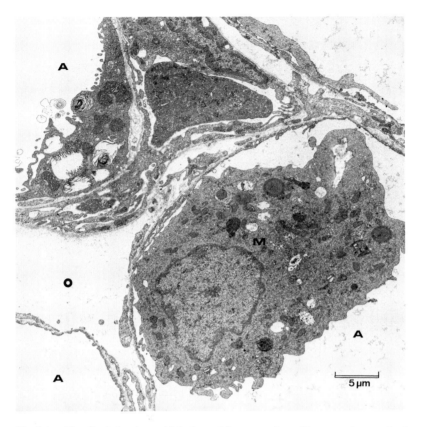

Fig. 3 A = Alveoli; circle = interstitial edema; M = macrophage. Electron micrograph of a typical interstional edema, which was often seen in the lungs of traumatized rats.

Results

In a previous study we were able to show that 48 h after drum shock and fracture of both thighs, the DPL content of the lung tissue was significantly reduced, compared to control values (fig. 1). Both controls and traumatized animals were nourished with 100 cal/kg body weight and day. In contrast to the marked reduction of the DPL content, the phospholipid content as well as the percentage of palmitic acid in the lecithin fatty acids, showed a slight reduction only.

In the present study the control rats were fed *ad libitum* and the traumatized animals were starved during this 48-hour period. In contrast to fed

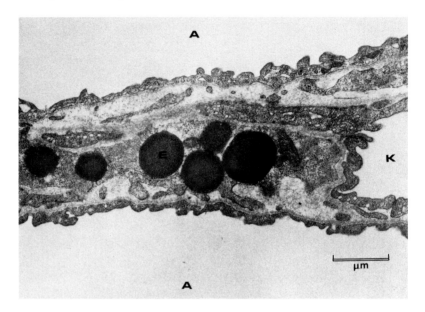

Fig. 4 A = Alveoli; E = osmiophilic inclusion bodies; K = capillary. Electron dense granules in the lung of a traumatized rat.

traumatized rats the phospholipid and DPL content in the lung tissue of the fasted traumatized rats was diminished. But the reduction of the DPL content was more pronounced. The ratio of DPL to total phospholipids decreased from 0.20, which is the control value, to 0.15 in the traumatized group (fig. 2).

The drum shock caused massive interstitial edema in the lung. Endothelial and epithelial cells (type I and II) were different to those of the control group (fig. 3). This contradicts earlier observations in dogs after traumatic shock, where some endothelial swelling and a massive accumulation of leukocytes were found. Very often some electron-dense granules could be detected in the lungs of the traumatized rats, but up to now we were unable to determine their significance (fig. 4). In general, no correlation between the PII cell structures in one-half of the lung, and the diminished DPL content of the other part of this lung could be found. However, there are no reliable facts which allow an exact delimitation of the amount of DPL built in the surfactant complex from the total amount present in the lung tissue. But it is possible that the marked reduction of DPL also results in a reduction of that amount of DPL which is available for the synthesis of the surfactant complex.

References

1   Bartlett, G.R.: Phosphorus assay in column chromatography. J. biol. Chem. *234:* 466–468 (1959).
2   Folch, J.: A simple method for the isolation and purification of the total lipids from animal tissues. J. biol. Chem. *226:* 497 (1957).
3   Lohninger, A. and Nikiforov, A.: Quantitative determination of natural dipalmitoyl lecithin with dimyristoyl lecithin as internal standard by capillary gas-liquid chromatography. J. Chromat. *192:* 185 (1980).

A. Lohninger, Forschungsinstitut für Traumatologie der Allgemeinen Unfall-Versicherungs-Anstalt, A-1200 Wien (Austria)

Discussion

Answering a question by *Clements, Lohninger* said that his very good separation of PC28–PC32 derivatives depends on using hydrogen rather than nitrogen as the carrier gas. He works with very high temperatures, 280–310°C, using commercial columns. *Van Golde* asked about the determination of the positions of fatty acids on lecithin. This was possible by using mass spectrometry. *Lohninger* added that in parenteral nutrition using lipid emulsions, with high percentages of linolic acid, the lung incorporates the linolic acid although there is no change in the DPL content. Confirming a question by *von Wichert, Lohninger* found that in the groups that received lipid emulsion there was an enhanced DPL content in the lung tissue of the shocked animals.

Prog. Resp. Res., vol. 15, pp. 224–233 (Karger, Basel 1981)

Correlation between Plasma Proteins and Surfactants[1]

C.M. Büsing, U. Bleyl and R. Rüfer

Institutes of Pathology and Pharmacology, Faculty of Clinical Medicine, Mannheim

Introduction

Intra-alveolar edema may have a low protein content (hydrostatic edema, e.g., in left ventricular failure) or be rich in protein (inflammatory or dyshoric edema, e.g., in pneumonia or in lesions of the alveolar wall). The difference is that in the normal hypophase and in hydrostatic edema there are only traces of plasma proteins [12]. In dyshoric, protein-rich edema a large amount of plasma proteins, e.g., albumin, globulin and fibrinogen are found [2].

Pathomorphology of Pulmonary Disturbances

In the lungs, the alveolar air space, the fluid-containing interstitium and the intravascularly circulating blood volume form a functional unit. A disturbance of one of these three components is followed by disturbances in the remaining compartments. In cardiogenic shock, for example, most prominent is the disturbance of peripheral microcirculation with tissue acidosis and activation of intravascular coagulation. The acid metabolic products and some vasoactive substances are flushed into pulmonary bloodstream. We know that disturbances of microcirculation also affect the terminal blood vessels of the lung [4, 14].

However, the pulmonary bloodstream has two special features: First, there are arteriovenular thoroughfare channels preceding the capillaries of

[1] Supported in part by a grant from the 'Deutsche Forschungsgemeinschaft', No. BU 415/1.

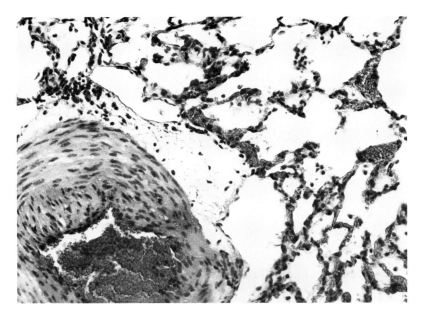

Fig. 1. Severe interstitial edema of the lung and lymphangiectasias. Formalin, paraffin, HE. × 195.

the lungs and with a lumen diameter varied by smooth muscle fibers [21]. Secondly, extensive interalveolar reticular capillaries are found in the lungs which rather resemble a capillary plate system in the alveolar walls, called the 'sheet-flow' system [9]. This capillary 'sheet-flow' is on the one hand favorable due to the large exchange areas for the blood gases, but on the other hand very sensitive to disturbances of the microcirculation.

Slight changes in the blood flow of the terminal pulmonary vessels rapidly lead to an erythrocyte aggregation with intense local elevation of viscosity: the blood no longer behaves as a Newtonian fluid [13, 15]. There thus arises pulmonary capillary stasis which is manifested morphologically as capillary ectasia. We term this 'congestive hyperemia'.

It should be mentioned that in a state of hypercoagulability these changes are intensified by the occurrence of intracapillary microthrombi which can be flushed into the pulmonary bloodstream or can arise locally.

A significant consequence of this disturbance of microcirculation is the impairment of the type II pneumocytes which are responsible for surfactant synthesis [7, 20]. The lack of the surface-active substances

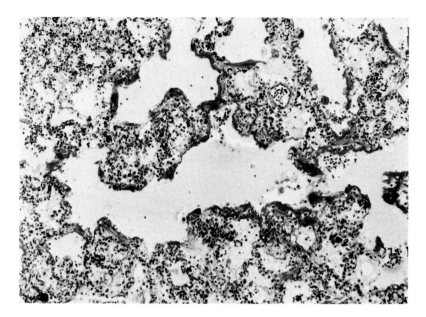

Fig. 2. Polytraumatization with respiratory distress syndrome: dystelectasias, protein-rich alveolar edema and pulmonary hyaline membranes. Formalin, paraffin, HE. × 195.

increases the negative pressure in the hypophase and in the interstitium. As long as the tight junctions of the alveolar epithelia (zonulae occludentes) are undamaged, an interstitial edema arises which can be drained to a certain extent via the lymphatic vessels which are enlarged (fig. 1) [8]. By pinocytosis and with further damage to the epithelia, fluid increasingly passes over into the hypophase [19]. In intravascular activation of coagulation, fibrin monomers and soluble oligomers necessarily pass into the alveolar lumina with the edema fluid, and then polymerize to produce the characteristic morphological picture of alveolar hyaline membranes (fig. 2) [4].

According to *Balis et al.* [1], the inhibition of surfactant by fibrinogen is irreversible. Consequently, there is a self-perpetuating relationship between intra-alveolar plasma protein and surfactant activity. The unphysiological content of plasma protein in the hypophase leads to an inactivation of the remaining surfactant. The surface tension thereby rises and the development of the intra-alveolar increases further. The increasing loss of intravascular fluid leads to intravascular hemoconcentration in the

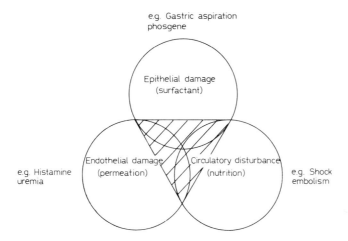

Fig. 3. Possibilities for initiation of the lung damage with development of a vicious circle. The shaded areas are the final features resulting in the lungs, which the pathologist will find and which are similar, irrespective of the primary injury.

lungs. The tendency of the erythrocytes to form aggregates increases, and the disturbance of microcirculation is again intensified.

At the same time, it can be inferred from this that besides the generalized imbalance of ventilation to perfusion, a different degree of inflation in the individual alveolus arises depending on how much surfactant is still active. There results the striking histological picture of dystelectasia with over-inflated alveoli next to partially collapsed alveoli. In addition, however, damage to the endothelium also becomes visible electron microsopically. In particular, permeability-active substances which are released from the thrombocytes, for example histamine, and further permeability factors may play a role [14].

The disturbance in one of the three parts of the functional unit of the lung also affects the other parts with the possibility of a vicious circle arising (fig. 3). A primary endothelial damage with disturbance of surfactant function is found, for example, in aspiration of gastric juice or in inhalation of toxic gases. A primary endothelial damage with increase of extravasation results, for example, in glomerulonephritis with uremia [3], or in the so-called bleomycin lung [6].

It can be seen form these interrelationships that there are many different possible ways in which the lung function may be disturbed.

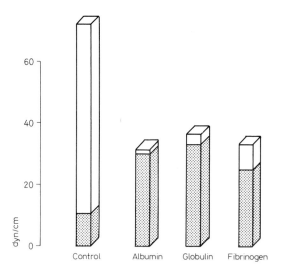

Fig. 4. Minimal surface tension diagram of 3 µg dipalmitoyllecithin on different hypo-phases (Ringer's solution) measured without and with protein (shaded area). Protein content: 10^{-12}g/100 ml [11].

However, one of the main mechanisms is impairment of the surfactant, followed by intra-alveolar extravasation of plasma (which itself impairs the surfactant activity).

Irrespective of the pathogenetic mechanism, the terminal morphological picture seen by the pathologist is the same. This is indicated by the shaded area in the diagram (fig. 3). It is only a question of time before all these lung lesions develop. Intensive care with artificial ventilation contributes to the process less by direct damage to the lungs than by affording the time for these changes to develop.

In vitro Inactivation of Surfactants

The inactivation of surfactant activity by plasma proteins has been demonstrated by several investigators [1, 17]. In *in vitro* investigations *Rüfer* [11] showed the striking impairment of surfactant function by each of the purified plasma protein components (fig. 4). We can apply this model to the lungs of patients with severe respiratory distress syndrome as well as to the experimental lung damage by oxygen.

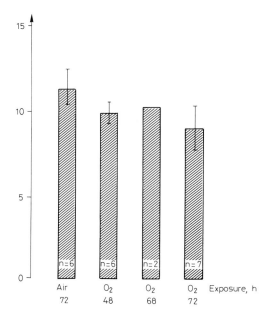

Fig. 5. Number of alveoli per visual field in controls and after 72 h of normobaric oxygen: no significant differences (Kruskal-Wallis test).

Experimental Findings

We have recently carried out experimental investigations on alterations in the lungs after exposure to normobaric hyperoxia in the rabbit. We have investigated these lungs histologically, morphologically and funtionally, giving special attention to protein-rich alveolar edema, to interstitial thickening and to the surfactant-dependent retraction of the lungs. We measured the latter by substraction of the retraction due to the tissue from the total retraction of the lung according to the method of *Rüfer* [10].

In the morphometric investigation, we measured the intra-alveolar area in each visual field. There was indeed a marked dyselectasia and the variation in the size of alveoli was striking. However, statistically we were not able to find a rise in the number of alveoli per unit area measured in the animals exposed to oxygen compared to the controls. This means that the collapsed alveoli are morpholigically compensated by the distended alveoli and the dilated terminal bronchioli (fig. 5).

On the other hand, there is a diminution in the total intra-alveolar area

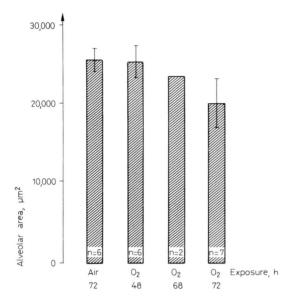

Fig. 6. Intra-alveolar area per visual field with significant diminution (p 0.05) after 72 h of normobaric oxygen exposure (Kruskal-Wallis test).

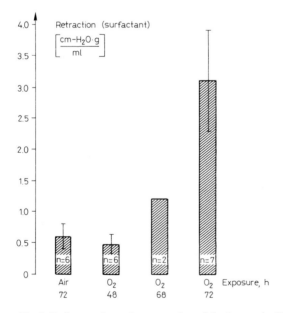

Fig. 7. Surfactant-dependent retraction of the lungs: significant increase (p 0.05) after 72 h of normobaric oxygen exposure (Kurskal-Wallis test).

Table I. Rabbit lung after normobaric oxygen exposure: histological findings after different times of exposure

	Number of animals			
	6	6	2	7
Alveolar edem.	0	0	0	7
PHM	0	0	0	3
Lymphangiektasia	0	2	2	6
Dystelektasis	3	4	0	7
Hyperemia	1	3	0	4
Exposure,	air	O_2	O_2	O_2
h	72	48	68	72

PHM = Pulmonary hyaline membranes.

measured. This results from an increase in the thickness of the interstitium. This alteration is not very striking, but is significant after 72 h exposure to oxygen (fig. 6). Our investigation of the function shows that the surfactant-dependent retraction of the lungs at first remains largely unchanged after oxygen exposure, but undergoes an abrupt increase after 72 h, that means there is a pronounced diminution of surfactant function here (fig. 7).

The histological investigation revealed that control animals do not show alveolar edema in any case. Even after exposure to oxygen for 48 h, there was no significant intra-alveolar accumulation of edema (table I). However, after 72 h all of the animals investigated displayed marked alveolar edema rich in protein. Hyaline membranes could be demonstrated in 3 cases. The lymphangiectasia of the lungs after oxygen exposure is striking. Lymphangiectasia could already be demonstrated after 48 h in 2 of the 6 animals and was noticeable after 72 h exposure to oxygen in 6 out of 7 animals. We think that it is the manifestation of an intensified lymph drainage of the interstitial fluid. One cause of the development of intra-alveolar edema may be that the drainage capacity of the lymphatic vessels is exceeded.

Summarizing these results and comparing them with the observations discussed above, it can be stated that lymphangiectasia and interstitial pulmonary edema after exposure to oxygen can be demonstrated relatively early, and that an alveolar edema then occurs, evidently with a time delay. According to *Weibel* [18], the impairment of the lung following normobaric hyperoxia starts initially with damage to the endothelia of the lung capillaries. On the one hand, this may lead to intravascular activation of coagulation [5] and on the other hand it seems to be the basis for the development of an interstitial edema. Recently, a direct impairment of the type II pneumocytes by normobaric oxygen could be demonstrated *in vitro* [16]. Besides this mechanism, the development of intravascular edema with plasma proteins in the hypophase is one of the central events in functional disorder of the lungs.

Summary

Alveolar air space, fluid-containing interstitium and blood volume in the intravascular circulation represent a functional unit in the lung. Disturbances may lead to a vicious circle because of the close connection between these three parts of the lung. One of the main events in these pathophysiological sequences is the interaction between the development of alveolar protein-rich edema and the impairment of the surfactants.

References

1   Balis, J.U.; Shelley, S.A.; McCue, M.J., and Rappaport, E.S.: Mechanism of damage to the lung surfactant system. Exp. molec. Path. *14:* 243 (1971).
2   Bleyl, U.: Pathomorphologie und Pathogenese des Atemnotsyndroms. Verh. dt. Ges. Path. *55:* 39 (1971).
3   Bleyl, U.; Werner, C. und Büsing, C.M.: Zur Pathogenese der uraemischen Pneumonitis. Klin.Wschr. *56:* 121 (1978).
4   Bleyl, U.: Die Histophysiologie und Histopathologie der terminalen Lungenstrombahn bei akutem Lungenversagen; in Ahnefeld, Bergmann, Buri, Dick, Halmagyi, Hossli und Rügheimer, Akutes Lungenversagen, pp. 1–13 (Springer, Berlin 1979).
5   Büsing, C.M. and Bleyl, U.: Oxygen induced pulmonary hyaline membranes (PHM) and disseminated intravascular coagulation (DIC). Virchows Arch. Abt. A. Path. Anat. Histol. *363:* 113 (1974).
6   Burckhardt, A.; Gebbers, J.-O. und Hölthe, W.-J.: Die Bleomycin-Lunge. Dt. med. Wschr. *102:* 281 (1977).
7   Goodwin, M.N., Jr.: Deficiency of pulmonary surfactant in metabolic acidosis. Am. J. Path. *62:* 49a (1971).
8   Lauweryns, J.M. and Baert, J.H.: Alveolar clearance and the role of the pulmonary lymphatics. Am. Rev. resp. Dis. *115:* 625 (1977).
9   Rosenquist, T.H.; Bernick, S.; Sobin, S.S., and Fung, Y.C.: The structure of the pulmonary interalveolar microvascular sheet. Microvasc. Res. *5:* 199 (1973).
10 Rüfer, R.: Der Einfluss oberflächenaktiver Substanzen auf Entfaltung und Retraktion isolierter Lungen. Pflüger's Arch. *298:* 170 (1967).

11 Rüfer, R.: Surfactant inhibition *in vitro.* XXV. Int. Congr. Physiological Sciences, Proc. Int. Union of Physiological Sciences IX 1971, abstract.

12 Schoedel, W.: Physiologische Grundlagen des Atemnotsyndroms. Verh. dt. Ges. Path. *55:* 2 (1971).

13 Schmid-Schönbein, H.: Zelluläre Physiologie der Mikrozirkulation: Ausbildung von Risikofaktoren als Folge optimaler Anpassungsfähigkeit; in Ahnefeld, Burri, Dick und Halmagyi, Mikrozirkulation, pp. 1–18 (Springer, Berlin 1974).

14 Schmutzler, W.: Humorale und immunologische Aspekte des akuten Lungenversagen; in Ahnefeld, Bergmann, Burri, Dick, Halmagyi, Hossli und Rügheimer, Akutes Lungenversagen, pp. 26–35 (Springer, Berlin 1979).

15 Sunder-Plassmann, L. und Messmer, K.: Funktionelle Veränderungen der Mikrozirkulation im Schock; in Ahnefeld, Burri, Dick und Halmagyi, Mikrozirkulation, pp. 76–86 (Springer, Berlin 1974).

16 Simon, I.M.; Raffin, T.A.; Douglas, W.; Theodore, J., and Robin, E.: Effects of high oxygen exposure on bioenergetics in isolated type II pneumocytes. J. appl. Physiol.: Resp. Envir. Physiol. *47:* 98 (1979).

17 Taylor, F.B. and Abrams, M.E.: Effect of surface active lipoprotein on clotting and fibrinolysis and of fibrinogen on surface tension of surface active lipoprotein. Am. J. Med. *40:* 346 (1966).

18 Weibel, E. Toxische Auswirkung erhöhter Sauerstoffspannung auf die Lunge; in Wiemers und Scholler, Lungenveränderung bei Langzeitbeatmung (Thieme, Stuttgart 1973).

19 v. Wichert, P.: Alveolarwandphysiologie und Surfactant. Verh. dt. Ges. Path. *62:* 29 (1978).

20 v. Wichert, P.; Wiegers, U.; Stephan, W.; Huck, A.; Eckert, P., and Riesner, K.: Altered metabolism of phospholipids in the lung of rats with peritonitis. Res. exp. Med. *172:* 223 (1978).

21 Zweifach, B.W.: Functional behavior of the microcirculation (Thomas, Springfield 1961).

Priv. Doz. Dr. C.M. Büsing, Institute of Pathology, Faculty of Clinical Medicine, Theodor-Kutzer-Ufer, D-6800 Mannheim (FRG)

Discussion

Reiss said that using oxygen would cause a lot of free radicals in the tissue, e.g., peroxides and similar compounds which can attack cellular structures. Therefore, when using oxygen one has to consider these effects besides those on the surfactant and he asked whether changes occurred not only in the plasma proteins but also in the plasma lipoproteins. *Rüfer,* talking about the role of proteins or other substances in the subphase said that these effects could be explained as an inhibition of surface spreading. Thus, the surface tension of the subphase can completely inhibit the spreading of surface-active substances. *Büsing* mentioned experiments showing that inhibition of the spreading induced by plasma proteins may be reversible but when induced by fibrinogen is not.

Therapeutical Approaches

Prog. Resp. Res., vol. 15, pp. 234–239 (Karger, Basel 1981)

Morphological Approach to Surfactant Secretion in the Lungs with Particular Reference to Ambroxol[1]

G. Elemer and Y. Kapanci

Department of Pathology, Faculty of Medicine, University of Geneva, Geneva

Indroduction

Earlier studies suggested that Bisolvon® (Boehringer) and more particularly its metabolite VIII (Na 872, i.e., Ambroxol) increases the pulmonary production of surfactant [*Curti,* 1972; *Lorenz et al.,* 1974; *Van Petten et al.,* 1978; *Prevost et al.,* 1979].

In our laboratories, it has been demonstrated that *per os* administration of Ambroxol is followed by an increase in type II epithelial cell volume and of lamellated body volume [*Cerutti and Kapanci,* 1979]; we thus postulated that this substance could increase type II epithelial cell activity.

The following autoradiographic investigations were done in order to appreciate the effect of Ambroxol on surfactant synthesis. Tritiated palmitic acid was administered to rats to study the uptake of this fatty acid by type II epithelial cells, and thus to estimate surfactant synthesis in the lung [*Askin and Kuhn,* 1971].

Material and Methods

27 Wistar rats weighing about 100 g were divided into three groups. Group I: rats received 200 mg/kg/day of Ambroxol *per os* during 3 days. Group II: rats were treated during the same period with 0.9% solution of NaCl. Group III: rats did not receive any treatment. The animals were fasted for 15 h. On the morning of the 4th day of the experiments, they were

[1] Supported by K. Thomae, Biberach. The morphometric equipment has been provided by SNSF, grant No. 3.382.78

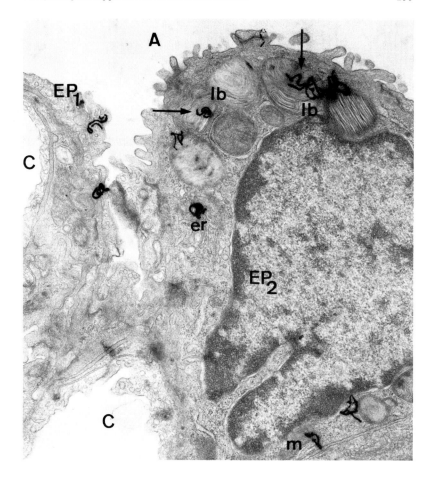

Fig. 1. In type II epithelial cell (EP₂) predominantly the lamellated bodies (lb) are labelled (grains = arrows). There are a few grains on the endoplasmic reticulum (er) and on the mitochondria (m). Note the labelling of type I epithelium (EP₁). A = Alveolus; C = capillaries. × 14.500.

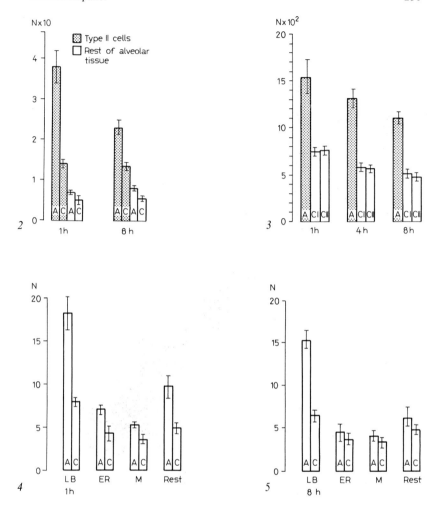

Fig. 2. Light microscopic evaluation (means and SE) of grain density in type II epithelial cells. A = Ambroxol-treated rats; C = controls; CI = 0.9 NaCl-treated rats; CII = untreated rats.

Fig. 3. Electron microscopic evaluation (means and SE) of grain density in type II cells (full columns) and in the rest of the alveolar tissue (empty columns). A = Ambroxol-treated rats; C = controls.

Fig. 4. Electron microscopic evaluation (means and SE) of grain density in the organelles of type II cells at 1 h. A = Ambroxol-treated; C = controls; LB = lamellated bodies; ER = endoplasmic reticulum; M = mitochondria.

Fig. 5. Same evaluations as in figure 4 but at 8 h.

injected with 9,10 palmitic acid ^3H (New England Nuclear Corporation, specific activity 23 Ci/mmol) complexed to bovine serum albumin as a potassium salt, in 0.9% sodium chloride. Injections were done into the jugular vein at a dose of 0.5 mCi/rat. 1, 4 or 8 h after injection, the rats were anesthesized with Nembutal (0.15 ml/100 g body weight) and the lungs were fixed in situ by perfusion as described previously [*Assimacopoulos and Kapanci,* 1974]. Small tissue blocs were cut from each lobe, washed in cacodylate buffer, postfixed in osmium tetroxide, and stained en bloc with uranyl acetate; they were dehydrated rapidly in 70 and 90% ethanol and embedded in Epon. Material for autoradiography was prepared according to the method of *Stein and Stein* [1967]. For light microscopic evaluation, 1-μm thick sections were cut on a Reichert ultratome, covered by an Ilford L-4 emulsion, exposed for 3 months, developed in a D 19 developer and slightly stained with methylene blue. For electron microscopic autoradiography, sections showing white interference colors were prepared and covered with an Ilford L-4 emulsion, exposed for 3 months, developed in Microdol X and investigated with a Philips EM 300 electron microscope.

The 'grain density' was evaluated by the light microscope at 800 times magnification under oil immersion. For electron microscopic studies, pictures were taken at 18,470 times magnification. Both by light and electron microscopy, the number of grains falling on a given structure was counted, and at the same time the respective volume density of the same structures was evaluated using the *Weibel* [1969] morphometric methods. The grain density was expressed as the number of grains per unit volume of tissue. This density was evaluated in type II cells and in the rest of the alveolar tissue as well as in the organelles of type II cells. The statistical evaluation of results was done using the Student's t test.

Results

Direct observations with the light microscope demonstrated that the grains were distributed all over the alveolar tissue but there was a manifest condensation within the type II epithelial cells. With the electron microscope, it was noted that the 'grains' predominantly labelled the lamellated bodies. However, a few were dispersed over the rest of the organelles, particularly on the endoplasmic reticulum (fig. 1). With these direct observations, no manifest difference could be detected between Ambroxol-treated and control animals.

Light microscopic morphometric evaluations, which give a very crude appreciation concerning the grain density, showed that at 1, 4 and 8 h the grain density was 2- to 3-fold higher in Ambroxol-treated rats than in controls (fig. 2). Furthermore, the grains were concentrated in the type II epithelial cells, their density being about ten times superior to that of the rest of the alveolar tissue.

Although less pronounced, the same type of distribution was observed by *electron microscopic countings* (fig. 3). The difference in grain density between Ambroxol-treated and control animals was highly significant (2p < 0.001). On the other hand, in Ambroxol-treated rats, there was an apparent drop between 1 and 8 h (fig. 3). and this difference was also significant ($2p < 0.05$). In control animals the grain density appeared to remain unchanged.

In type II cells of treated and control animals, the highest degree of labelling occurred in lamellated bodies (fig. 4, 5). In these structures, the grain density was more than 2-fold higher in Ambroxol-treated rats than in controls. Both at 1 and 8 h, the difference was highly significant ($2p < 0.001$). Between 1 and 8 h, an insignificant drop occurred in the labelling of the lamellated bodies of experimental and control rats (fig. 4, 5).

At 1 h, not only the lamellated bodies but also other organelles of the type II cells had a higher grain density in experimental animals (fig. 4). The difference was highly significant for 'the rest' of the cell comprising the cell membrane and surface villosities. It was also significant for the endoplasmic reticulum and Golgi area ($2p < 0.05$) but not for the mitochondria. At 8 h, except for the lamellated bodies, no significant difference in grain density was noted in the organelles of type II cells (fig. 5).

Comment

Our findings demonstrate that in Ambroxol-treated animals, the incorporation of tritiated (^{3}H)-palmitic acid into the alveolar tissue occurs at a much higher rate than in controls. This incorporation takes place at the level of the type II epithelial cells and particularly into the lamellated bodies. Palmitic acid is a constituent of dipalmitoye lecithin which, in turn, is a major constituent of the pulmonary surfactant [*Young and Tierney*, 1972]. Hence an increase in its incorporation into the lamellated bodies indicates an increase in the synthesis of the pulmonary surfactant. The fact that at 1 h the labelling of the endoplasmic reticulum is also more pronounced is probably a further argument in favor of an increased surfactant synthesis in the lung of Ambroxol-treated rats.

Why, at 8 h, the grain density in type II cells as well as in the lamellated bodies tends to decrease is not yet fully understood. It is possible that by then the surfactant synthetized by incorporation of (^{3}H)-palmitic acid is secreted into the alveoli. However, further biochemical analyses are required in order to answer the question whether the increase in synthesis of the surfactant also implies an increase in its secretion. These studies are now underway in our laboratories.

References

1 Assimacopoulos, A. and Kapanci, Y.: Lung fixation by perfusion. A simple method to control the pressure in perfusion circuit. J. Microsc., Oxford *100:* 227–229 (1974).
2 Askin, F.B. and Kuhn, C.: The cellular origin of pulmonary surfactant. Lab. Invest. *25:* 260–268 (1971).
3 Cerutti, P. and Kapanci, Y.: Effects of metabolite VIII of Bromexine (Na 872) on type II epithelium of the lung. Respiration *37:* 241–251 (1979).
4 Curti, P.C.: Steigerung der Produktion der oberflächenaktiven Substanz der Alveolen durch Na 872. Pneumologie *142:* 62–74 (1972).
5 Lorenz, U.; Ruttgers, H.; Fux, G., and Kubli, F.: Fetal pulmonary surfactant induction by bromexine metabolite VIII. Am. J. Obstet. Gynec. *119:* 1126–1128 (1974).
6 Prevost, M.C.; Soula, G., and Dousle-Blazy, L.: Biochemical modifications of pulmonary surfactant after bromexine derivate injection. Respiration *37:* 215–219 (1979).
7 Stein, O. and Stein, Y.: Lipid synthesis, intracellular transport storage and secretion. I. Electron microscopic radioautographic study of liver after injection of tritiated palmitate or glycerol in fasted and ethanol-treated rats. J. Cell Biol. *33:* 319–339 (1967).
8 Van Petten, G.R.; Mears, G.J., and Taylor, P.: The effects of Na 872 on pulmonary maturation in the fetal lamb and rabbit. Am. J. Obstet. Gynec. *130:* 35–40 (1978).
9 Weibel, E.R.: Stereological principles for morphometry in electron microscopic cytology. Int. Rev. Cytol. *26:* 235–302 (1969).
10 Young, S.L. and Tierney, D.F.: Diplamitoyl lecithin secretion and metabolism by the rat lung. Am. J. Physiol. *222:* 1539–1544 (1972).

Dr. G. Elemer Department of Pathology, 40, boulevard de la Cluse, CH-1211 Geneva 4 (Switzerland)

Discussion

Experiments showing that Ambroxol significantly elevates the synthesis of phosphatidylcholine and phosphatidylglycerol from a number of precursors were reported in the discussion of *van Golde. Geiger Mannheim* pointed out that using autoradiography one should be very careful because the energy spectrum of the substances used may lead to contradictory results. *Reiss* mentioned experiments performed in collaboration with *Enhorning,* using Ambroxol in baby rabbits; when given to the mothers prior to birth, the lipid phosphorous does not change in lavage material, but a very strong protein decrease in the lung lavage was found. *Ellema* pointed to the difference in ages of the experimental animals and *Kapanci* mentioned experiments in addition to earlier experiments from *Weibel* that there is also an increase in lamellar body volume and number of type II cells of 200% with Ambroxol. Answering a question by *Zänker, Ellema* said that it was an increase in grain density over the type II cells which she had observed rather than an increase of the number of type II cells. There is an increase in cell volume but not in cell number.

Prog. Resp. Res., vol. 15, pp. 240–255 (Karger, Basel 1981)

Effects of Ambroxol (NA 872) on Biochemical Fetal Lung Maturity and Prevention of the Respiratory Distress Syndrome

E. Löwenberg, L. Jiménez, M. Martínez and M. Pommier

20th of November Medical Center ISSSTE, México

Introduction

The respiratory distress syndrom (RDS) is the most important disorder caused by a deficiency of lung surfactant activity [1, 6]. Prenatal diagnosis may be made through analysis of the phospholipids in the amniotic fluid [8,9,11,20,21]. Lung maturation acceleration in the fetus through prenatal administration of glucocorticoids has been a controversial subject [2,4,10,16–18,23]. The usefulness of other substances has not been sufficiently explored [14,24]. Studies made with a bromhexine metabolite, Ambroxol, have recently attracted the attention of researchers in this field.

Chemically, Ambroxol (NA 872) is a trans-4 [2-amino-3,5-dibromobenzyl, amino] cyclohexanol-hydrochloride. Pharmacological studies on animals disclosed that the compound has a stimulating action on bronchial secretions, increases the alveolar surface-active substances, probably due to proliferation of type II alveolar cells, and increases the number of osmophil bodies [7,13,16]. Similar effects were found in tests performed on lung epithelium cell growth [26]. Comparing volume-pressure diagrams of the lungs of mice treated with Ambroxol and of a control group, it was found that the former were collapse-resistant [3]. Other studies, however, showed Ambroxol to have no influence on the synthesis of surfactant substances in animal experiments [16]. The elimination half-life of [14] C-labelled compound radioactivity in blood was estimated as 20–25 h in rat, dog and man [12]. Toxicity is very low, even when given in high doses. Studies made with Ambroxol on pregnant women showed a stimulating effect on fetal surfactant substances, which

Table I. Diagnosis made on admission of patient

Diagnosis	Treated	Control
Threatened preterm labor	40	33
Isoimmunization	1	6
Diabetes	0	2
Hypertension	1	4
Urosepsis	4	6
IUGR	0	3
Bad fetal conditions	2	4
Combination of the two preceding conditions	2	3
Premature rupture of membranes	10	4
Toxemia	9	5
Total	70	69

IUGR = intrauterine Growth Retardation

was not always commensurate with a decrease in RDS incidence in the newborn babies [5,19,25].

This study was designed to determine the effects of Ambroxol (NA 872) on women with threatened preterm labor or women who required, for medical reasons, interruption of gestation between the 30th and 35th weeks, evaluation of phospholipid modifications in the amniotic fluid, and the possible abatement of RDS incidence in newborn babies. Furthermore, it was sought to determine the optimum dosage required to obtain satisfactory results.

Material and Method

At the beginning the group comprised 85 treated patients, but only 70 fulfilled the requirments for being included in this study. In addition there was a control group of 69 patients. Admittance diagnoses of both groups are shown in table I. Number of pregnancies and gestational age for the group under treatment as well as for the control group are given in tables II and III.

The group under treatment included patients with threatened preterm labor and patients who were advised to terminate pregnancy for medical reasons. The control group was formed by 69 patients who arrived in preterm labor conditions and from whom amniotic fluid was obtained for test purposes. They had not received fetal lung maturity induction treatment, since all gave birth within a week from which amniocentesis was performed. Utero-inhibition was attempted in the treated group as well as in the control group with betamimetic drugs.

Table II. Number of previous pregnancies in groups studied

Previous pregnancies	Control	Treated
First pregnancy	28	18
2–4 pregnancies	12	30
More than 5 pregnancies	29	22
Total	69	70

Table III. Gestational age at trial start

Weeks	Control	Treated
30	5	5
31	4	4
32	7	14
33	9	9
34	11	19
35	33	19
Total	69	70

The treated group was handled as follows: patients admitted in an impending stage of labor were sent directly to the Perinatal Medicine Service in order to immediately start a utero-inhibition procedure after tocological evaluation. Terbutaline or fenoterol were then used in the required amounts and manner (5 µg/min and 1 or 2 µg/min with an infusion pump, respectively). A transabdominal amniocentesis was performed to determine the phospholipid content of the amniotic fluid with the technic described by *Hallman and Gluck* [11]. 1,000 mg of Ambroxol was administered, diluted in 250 ml of a 5% glucose solution with the aid of an infusion pump adjusted for a 2-hour drip. If the patient did not give birth, the treatment was repeated every day until 5 doses were completed. Then it was stopped to be resumed later with five weekly administrations extending into the 35th week, when the treatment was stopped altogether. The majority of the patients gave birth before the full treatment was completed, or did not attend to complete their doses.

When pregnancy was interrupted for obstetrical reasons a diagnostic amniocentesis was performed and fetal lung maturation was stimulated by administration of 1,000 mg of Ambroxol diluted in 250 ml of a 5% glucose solution in a 2-hour drip. Treatment was repeated daily until pregnancy was interrupted. A sample of the amniotic fluid was taken at the end of the treatment or during obstetrical handling.

This report only includes cases where amniotic fluid was obtained for later examination either through transabdominal amniocentesis or through vaginal fluid collection. The figures of the report refer only to the first treatment, since it was not always possible to perform repeated amniocentesis after each treatment.

As some patients gave birth at different gestational ages and received different numbers of dosages, we did not attempt to form additional groups according to doses received as it would mean breaking up the whole group into small, insignificant groups.

The following reference levels were established: a lecithin/sphingomyelin (L/S) ratio equal to or lower than 2 with negative phosphoglycerol (L/S$\leq$2, PG negative) was established as the fetal lung immaturity level, and an L/S ratio above 2 with positive phosphoglycerol L/S$\geq$2, PG Positive) as the lung maturity level where the possibility of RDS is small.

The increase in the L/S ratio after treatment was compared with the results obtained from the control group as well as with the L/S ratio before administration of Ambroxol. In order to compare the results obtained during the different weeks of pregnancy, considering that changes in fetal lung maturity are influenced by gestational age at the time of examination, the interval between one amniocentesis and the next, as well as gestational pathology, it was considered advisable to design a biochemical lung maturity index or conversion maturity index (CI), based on the number of cases in which biochemical lung immaturity (L/S$\leq$2, PG negative) turned to positive (L/S$\geq$2, PG positive) after treatment multiplied by 100. Natural occurrence of CI from the 30th to 35th weeks of gestational age in the control groups was recorded.

Statistical analysis of L/S ratio and CI figures before and after treatment were compared with control groups (linear regression and Student's t test).

Side effects of the drug were recorded. Cardiotocographic control was performed during administration of Ambroxol and patients receiving the full treatment were kept under biweekly observation until childbirth. 1-min and 5-min Apgar scores, weight, presence or absence of RDS, as well as mortality rate due to RDS were recorded.

Results

Side Effects

210 doses were used in 70 patients. Nausea and vomiting were fairly common during Ambroxol infusion, the former in 44 doses applied to 21 patients and the latter in 15 doses applied to 9 patients (21 and 7% of total number of doses, respectively). In 4 of them it was necessary to stop the treatment temporarily. The same effects were reported in some of the patients of the control group who received utero-inhibitor drugs.

Most of the mothers and fetuses presented tachycardia due to the use of betamimetic drugs in the treated as well as in the control group. There was slight tachycardia (5%) in those cases where the only drug used was Ambroxol. No significant blood pressure alterations were observed. Cardiotocographic records did not show any significant changes. Apgar scores showed no difference between the babies in the treated and control groups. Weight of the babies in both groups was within normal limits for similar gestational ages and there were only 8 cases of hypotrophic babies in the

Table IV. Figures for cases treated with NA 872 (Ambroxol) and control group in weeks 30 and 31

Total number		n	x̄ L/S	s	L/S≤2 PG-	L/S>2 PG+	CI	Morbidity RDS	Mortality RDS	Weight x̄, g	Doses NA 872 1	2	3	4	5	Week of delivery
Week 30																
Control group	5	RDS	4	1.00	0.56	4	0		4	2	1,080						
		no RDS	1	1.22		1	0	0			1,120						30, 31
Treated group	5	before NA 872	5	0.94	0.26	5	0						2		1		30, 31+ \ 30
		after NA 872	5	1.73	0.16	5	0	0	1	1	1,434				1		33 \ 36
Week 31																	
Control group	4	RDS	1	1.20		1	0		1	1	1,150						31, 32
		no RDS	3	1.37	0.068	3	0	0			1,298						
Treated group	4	before NA 872	4	1.33	0.75	4	0	¼	1			1					31+
		after NA 872	4	2.00	0.49	3	1	25	1	1	1,595		1		1	1	32 \ 34 \ 36

s = standard deviation

control group, against 5 in the treated group. Phospholipid changes and RDS cases are reported separately for each gestational week.

30th Week

Control Group. The control group included 5 patients. The average L/S ratio was 1.00 for RDS babies and 1.22 for babies without this disorder as reported in table IV. 4 newborns developed RDS, with 2 deaths and 2 survivals.

1 did not develop RDS and his development was satisfactory. The average weight and gestational age at delivery of the newborns are reported in table IV. All the 5 newborns presented data of biochemical lung immaturity, so CI was 0.

Treated group. The group included 5 cases who were under treatment for 1–5 consecutive days. All of them showed an improvement in the L/S ratio (average 0.94 before and 1.72 after treatment). There was 1 case with a 100% L/S ratio increase after 5 doses (from 0.80 to 1.64). Utero-inhibition was successfully accomplished in this patient who received 3 complete 5-dose treatments at weekly intervals and gave birth in the 36th week. In another case the patient gave birth after a single dose to a female baby weighing 1,050 g, Apgar score 4–6, who developed RDS and died 48 h later. 3 pregnant women gave birth between the 30th and the 31st week; 1 in the 33rd week and 1 in the 36th week. 3 cases gave birth after 1, 2 and 3 doses, respectively; the babies did not develop RDS. L/S ratio increased by between 10 and 100% in all 3 cases. Clinical evolution of the babies was commensurate with normal preterm babies weighing between 1,100 and 1,290 g.

All 5 cases showed negative lung maturity index at the beginning of treatment, which remained unchanged throughout. CI was 0. Statistical analysis of the average L/S ratio before and after treatment was highly significant; not so on comparison with the control and treated groups' figures found in the 31st week, when the second amniocentesis was performed.

31st Week

Control Group. This group included 4 cases. 1 developed RDS and died on the 7th day. L/S ratio was 1.20 for the baby developing RDS and an average of 1.37 for those who did not present RDS. Figures are shown in table IV. CI was 0. The baby who died weighed 1,150 g and the other 3 were on average 1,298 g.

Table V. Figures for cases treated with NA 872 (Ambroxol) and control group in weeks 32 and 33

Total number		n	x̄ L/S	s	L/S ≤ 2 PG−	L/S > 2 PG+	CI	Morbidity RDS	Mortality RDS	Weight x̄, g	Doses NA 872 1 2 3 4 5	Week of delivery
Week 32												
Control group 7	RDS	2	1.38	0.080	2	0	$1/7$ 2		1	1,250		
	no RDS	5	1.45	0.741	4	1	14			1,280		32
Treated group 14	before NA 872	13	1.64	0.296	11		$4/11$ 0				5	32, 33, 34, 35, 36+
											2	32, 34
										1,700	3	32, 32, 40 RDS
	after NA 872	14	2.18	0.428	7	4	36 2		1		2	32, 38
											2	33, 37
Week 33												
Control group 9	RDS	3	1.40	0.082	3	0	$2/8$ 3		1	1,580		
	no RDS	6	1.74	0.164	3	2	25			1,710		33
Treated group 9	before NA 872	9	1.93	0.180	7	0	$4/7$				1	33
											4	33, 33, 35, 37+
	after NA 872	9	2.24	0.47	3	4	57 1		1	2,300	2	33, 33
											1	36
											1	38

Treated Group. Of the 4 cases included in this group, only 1 received the full 5 doses with a 100% increase in the L/S ratio. In the remaining cases, the increase in the L/S ratio was from 10 to 60% (all 4 cases increased from 1.33 before to 2.00 after Ambroxol). 1 of the cases developed RDS. After a single dose of Ambroxol the L/S ratio increased from 1.25 to 1.40. The baby's weight was 1,000 g. It died 48 h later. The average weight of the babies in this group was 1,595 g. 1 of the patients gave birth in the 36th week. 1 case received 3 daily doses and turned lung maturity data from negative to positive. CI was 25. No statistical significance was found between figures before and after treatement.

32nd Week

Control Group. 7 patients were studied in this group. 2 of the newborn babies developed RDS; 1 of them died, the other survived. The average weight in the first subgroup was 1,250 g. 5 babies did not present RDS, 1 of them had figures of biochemical lung maturation, average weight in this subgroup was 1,280 g. CI was 14 (1 out of 7 presented L/S $\geqslant$ 2, PG positive). Details on phospholipids analysis are shown in table V.

Treated Group. The treated group included 14 patients. Initial amniocentesis was unsuccessful in 1 case. Average L/S ratio at the beginning was 1.64 and after treatment 2.18. 11 out of 13 amniocenteses showed figures of lung immaturity (L/S $\leqslant$ 2, PG, negative); after Ambroxol 7 remained negative and 4 turned positive. 2 had an L/S $\geqslant$ 2, PG negative and did not fulfil the requirements necessary for the evaluation of CI. CI was 36 (4 out of 11).

There were 2 RDS cases; 1 received 3 doses and survived and the other received 1 dose an died. Utero-inhibition was successful in 3 cases. 5 patients gave birth shortly after the last Ambroxol administration. 2 of the cases of successful conversation maturity index had received 1 dose, the remaining cases received 2 and 4 doses. The average weight of the newborn babies was 1,700 g (table V).

L/S ratio figures after treatment were statistically significant compared with the control group and before-treatment figures (p values less than 0.05). They were also significant when compared with the control group figures of the 33rd week, in which the second amniocentesis was performed.

Table VI. Figures for cases treated with NA 872 (Ambroxol) and control group in weeks 34 and 35

		Total number	n	x̄ L/S	s	L/S≤2 PG−	L/S>2 PG+	CI	Morbidity RDS	Mortality RDS	Weight x̄, g	Doses NA 872 1	2	3	4	5	Week of delivery
Week 34																	
Control group	RDS	11	3	1.65	0.420	3	0	4/11	3	0	1,810						
	no RDS		8	1.95	0.361	4	4	36			1,970						34
Treated group	before NA 872	19	19	1.90	0.254	11	0	6/11					2		4		30, 35 34, 34, 36, 36 34, 35, 38
	after NA 872		19	2.37	0.338	5	6	55	0	0	2,620			3		5	34, 34, 35, 37, 40 34, 36, 38, 40, 40
Week 35																	
Control group	RDS	33	5	2.10	0.462	5	0	15/33	5	1	2,320						35
	no RDS		28	2.30	0.370	13	15	45			2,450						
Treated group	before NA 872	19	19	2.24	0.260	16	0	10/16						3	4		35, 35, 36 35, 35, 36, 37 35
	after NA 872		19	2.68	0.206	6	10	63	0	0	2,680		2			5	35, 35, 36, 37, 40 35, 37, 38, 40, 40

33rd Week

Control Group. This group included 9 patients. 3 of the newborn babies developed RDS; 1 of them died. The average L/S ratio in RDS cases was 1.40. The 6 RDS-free cases showed an average L/S ratio of 1.74.

In the control group there were 2 cases with L/S $\geqslant$ 2, PG positive as well as 6 L/S $\leqslant$ 2, PG negative cases. There was also 1 case with L/S $\geqslant$ 2 and PG negative; CI was 25 (table V). The average weight was 1,580 g in the first subgroup and 1,710 g in the second.

*Treated Group.*9 cases were treated in this group. Before treatment the average L/S was 1.93, it increased to 2.24 after treatment. At the onset there were 7 cases with L/S $\leqslant$ 2, PG negative with 4 positive conversions after treatment, or a CI of 57. L/S $\geqslant$ 2, PG negative was found in 2 cases, converted into positive PG in 1 of the cases after administration of 3 doses (table V). In this group, as in the preceding ones, there was a minimum rate of completed treatments. The majority of cases received only 2–3 doses (6 cases).

The average weight of the newborn babies was 2,300 g. In 2 cases, pregnancy continued beyond the 37th week. There was one death due to RDS, the mother received 2 Ambroxol doses after which she gave birth. L/S ratio values obtained before treatment were significant compared with the figures after Ambroxol treatment, but not significant compared with the values of the 34th week, when the second amniocentesis was performed.

34th Week

*Control Group.*This group included 11 patients, there were 3 RDS cases and an average L/S ratio of 1.65; 2 of them were classified as transitory RDS and returned to normal after 24 h; the other survived after a 6-day treatment. Average weight in this subgroup was 1,810 g. 8 pregnancies ended with babies weighing an average of 1,970 g and did not develop respiratory failures. 7 out of 11 showed an L/S $\leqslant$ 2, PG negative and 4 an L/S $\geqslant$ 2, PG positive. CI was 36 (table VI).

*Treated Group.*There were 19 cases under treatment, only 5 cases completed a 5-dose treatment with an average L/S ratio which increased from 1.76 before to 2.39 after treatment. For the entire group the L/S ratio average was 1.90 before the Ambroxol treatment which increased to 2.37

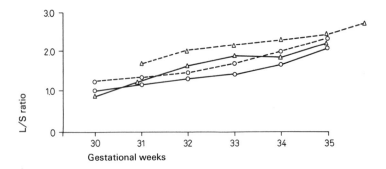

Fig. 1. Influence of Ambroxol on lecithin/sphingomyelin (L/S) ratio at start and during treatment. Control group: O—O = RDS; O---O = no RDS; treated group: △—△ = before Ambroxol △---△ = after Ambroxol.

after treatment. Statistical analysis was significant, but when compared with the control group figures of the 35th week it was not significant.

Fortunately, there was not a single case of RDS. There were 6 successful cases of premature labor inhibition with consequent full-term births. 8 of the cases showed an L/S ⩾ 2 with PG negative (table VI). From 11 cases which showed immaturity figures, 6 turned to positive after treatment with a CI of 55.

35th Week
Control Group. In 33 patients there were 5 newborn babies who developed RDS with an average L/S ratio of 2.10; all of them had an L/S ⩽ 2, PG negative. 1 of the babies died. The average weight in this group was 2,320 g. 28 newborn did not present RDS, 13 had an L/S ⩽ 2, PG negative and 15 with an L/S ⩾ 2, PG positive. Average weight was 2,450 g. CI was 45 (table VI).

Treated Group. 19 patients received Ambroxol, doses ranging from 1 (3 cases) to 5 (5 cases). The L/S ratio was 2.24 before treatment against 2.68 thereafter. 16 fluid samples showed an L/S ⩽ 2, PG negative at the onset of treatment, successfully converted to positive in 10 cases after treatment. 6 cases continued to be negative (5 with an L/S ⩾ 2, PG negative). Although the L/S ratio was increased successfully, the cases were still rated as negative because they did not conform to the positive type standard determined before the test. Not a single case of RDS was encountered. The

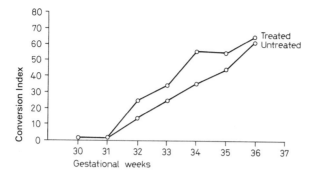

Fig. 2. Influence of Ambroxol on L/S ratio and presence of phosphatidyl glycerol (conversion index, CI) during pregnancy.

$$CI = \frac{L/S\ >2,\ PG\ \ positive}{L/S\ \leqslant 2,\ PG\ \ negative} \times 100.$$

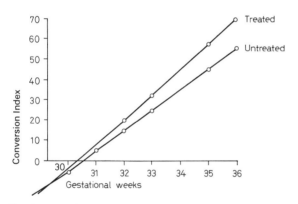

Fig. 3. Regression analysis of CI in group treated with Ambroxol and in the untreated group.

average weight was 2,680 g. Pregnancy continued beyond the 37th week in 6 of the cases. 3 cases showed an L/S ⩾ 2, PG negative, and after treatment the L/S ratio increased, but PG remained negative. CI was 63. Details are given in tables IV–VI and figures 1 and 2.

There was no statistical difference in the weekly average L/S ratio data of the control group and the figures obtained before treatment. Statistical analysis of the data before treatment as well as figures with no RDS of the

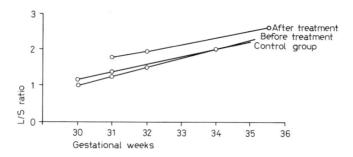

Fig. 4. Effect of Ambroxol. Regression analysis of the L/S ratio in the test group before and after treatment with Ambroxol and in the untreated group.

control group compared with those obtained after Ambroxol showed that they were significant (p values less than 0.05) in all the weeks except for the 31st week. 18 babies developed RDS in the control group against 5 in the treated group. 6 babies died in the former group and 4 in the latter. Comparison of the L/S ratio obtained after treatment with control group results from that week when the second sample was taken was statistically only significant (p<0.05) in the 30th and 32nd week.

Linear regression analysis was performed with CI of the controls as well as with the values of the treated group. The same treatment was given to the L/S ratio values of both groups. They were submitted to a statistical analysis and were found significant (fig. 3, 4).

Comments

After fetal lung maturation acelleration in sheep through the use of glucocorticoids [17], several research projects have been performed to determine the advantages and disadvantages of the use of the drug in humans. Findings have been controversial, for although there is a considerable proportion of successful cases, we cannot dismiss the cautionary statements of some authors who claim that there is a danger of long-term adverse effects on the newborn babies. Concern about this situation has prompted researchers to investigate other drugs with fewer undesirable effects and greater favorable action on the lungs.

Ambroxol (NA 872), the drug examined by us, is a bromhexine metabolite which has experimentally proven its effectiveness as a lung surfactant inducer [3,5,16]. Earlier studies have proven its effectiveness in humans. The test performed by us showed a significant increase in the L/S ratio after administration of the drug, although, on comparing the rates obtained in consecutive weeks, they proved to be negligible on weeks 31, 33 and 34. Although our lung maturity conversion index may be construed as a hopeful figure, it is based on substantiation provided by our earlier research work carried out on pregnancies resulting in the birth of RDS-free babies. The present data confirm that Ambroxol increases the conversion rate to a considerable extent. Statistical analysis was highly significant, further confirming the reliability of the conversion index method for evaluation of the drug. There were fewer babies with RDS in the treated than in the control group, 5:18. This would indicate that RDS decreases significantly under Ambroxol treatment. The number of pregnancies going beyond the 36th week was 38. The remaining cases gave birth before term. The cases were not classified according to previous pathology as this would have required an excessive break up of the already limited number of cases.

Interpreting the facts, it should be borne in mind that time as well as possible maternal pathology influence fetal lung maturity, also that uterine activity was present in some cases and absent in others, uterine activity increases and/or releases phospholipids into amniotic fluid.

It was not possible to determine the optimum dosage, but we believe that the L/S ratio increases after administration of 1 or more doses, while the conversion index appears more frequently on completion of 4–5 doses. The effects of utero-inhibitors should not be overlooked since, as stated at the beginning of this report, these drugs seem to stimulate the lung enzyme systems responsible for the increase in surface-active substances. In the control group they were only used for a short time, and in the treated group freely during Ambroxol treatment.

In view of the drug's satisfactory tolerance level, it would seem highly advisable to further extend research work to determine the possible immediate and long-term adverse effects on the newborn babies.

References

1   Brown, B.J.; Gorbert, H.A., and Stenchever, M.A.: Respiratory distress syndrome, surfactant biochemistry and acceleration of fetal lung maturity. A review. Obstetl gynec. Surv. *30:* 71–73 (1975).
2   Caspi, E.; Schreyer, P.; Weinraub, S.; Bukovsky, I., and Tamir, I.: Changes in amniotic

fluid lecithin/sphingomyelin ratio following maternal dexamethasone administration. Am. J. Obstet. Gynec. *122:* 327–331 (1975)

3   Curti, P.C.: Steigerung der Produktion der oberflächenaktiven Substanz der Alveolen durch NA 872. Pneumologie *147:* 62–74 (1972).

4   De Lemos, R.A.; Shermata, D.; Knelson, J.H.; Kotas, R.V., and Avery, M.E.: Acceleration of appearance of pulmonary surfactant in the fetal lamb by administration of corticoids. Am. Rev. resp. Dis. *102:* 358–361 (1971).

5   Agberts, J.; Fontijne, K., and Wamsteker, J.: Indication of increase of the lecithin/sphingomyelin (L/S) ratio in lung fluid of lambs maternally treated with metabolite VIII of Bisolvon. Biol. Neonate *29:* 315–322 (1976).

6   Farrell, P.M. and Avery, M.E.: Hyaline membrane diseases. Am. Rev. resp. Dis. *111:* 657 (1975).

7   Gil, J. and Thurnherr, U.: Morphometric evaluation of ultrastructural changes in type II alveolar cells of rat lung produced by bromhexine. Respiration *28:* 438–456 (1971).

8   Gluck, L.; Kulovich, M.; Borer, R.C.; Brenner, P.H.; Anderson, C., and Spellacy, W.N.: Diagnosis of the respiratory distress syndrome by amniocentesis. Am. J. Obstet. Gynec. *109:* 440–445 (1971).

9   Gluck, L. and Kulovich, M.V.: Lecithin/sphingomyelin ratios in amniotic fluid in normal and abnormal pregnancies. Am. J. Obstet. Gynec. *115:* 539–546 (1973).

10   Gluck, L.: Administration of corticosteroid to induce maturation of fetal lung. Am. J. Dis. Child. *130:* 976–978 (1976).

11   Hallman, M. and Gluck, L.: Development of the fetal lung. J. perinatal Med. *5:* 3–31 (1977).

12   Hammer, R.; Bozler, G.; Jauch, R. und Koss, F.W.: Speziesvergleich in Pharmakokinetik und Metabolismus von NA 872 Cl Ambroxol bei Ratten, Kaninchen, Hund und Mensch. Arzneimittel-Forschung Drug Res. *5a:* 899–903 (1978).

13   Kapanci, Y. and Cerutti, P.: Morphometric evaluation of type II epithelial-cell changes in the lungs of rats, produced by a metabolite of bromhexine (unpublished).

14   Kero, P.; Hirvonen, T., and Valimaki, I.: Prenatal and postnatal isoxsuprine and respiratory distress syndrome. Lancet *ü:* 198 (1973).

15   Kotas, R.V. and Avery, M.E.: Accelerated appearance of pulmonary surfactant in the fetal rabbit. J. appl. Physiol. *30:* 358–361 (1971).

16   Krieglsteiner, P.; Lohninger, A.; Munnich, W.; Erhard, W.; Neiss, A. und Blumel, G.: Effekt von Betamethason und Bromhexin-Metabolit VIII auf die fetale Surfactantbiosynthese (unpublished).

17   Liggins, G.C.: Premature delivery of foetal lambs infused with glucocorticoids. J. Endocr. *45:* 515–523 (1969).

18   Liggins, G.C. and Howie, R.N.: A controlled trial of *ante partum* glucocorticoid treatment for prevention of the respiratory distress syndrome in premature infants. Pediatrics, Springfield *50:* 515–525 (1972).

19   Lorenz, U.; Ruttgers, H.; Fux, G., and Kubli, F.: Fetal pulmonary surfactant induction by bromhexine metabolite VIII. Am. J. Obstet. Gynec. *119:* 1126–1128 (1974).

20   Lorenz, U.; Ruttgers, H.; Fromme, M., and Kubli, F.: Significance of amniotic fluid phospholipid determination for the prediction of neonatal respiratory distress syndrome. Z. Geburtsh. Perinat. *179:* 101–111 (1975).

21   Nelson, G.H.: Relationship between amniotic fluid lecithin concentration and respiratory distress syndrome. Am. J. Obstet. Gynec. *112:* 827–833 (1972).

22   Puschmann, S. und Engelhorn, R.: Pharmakologische Untersuchungen des Bromhexin-Metaboliten Ambroxol. Arzneimittel-Forsch. (Drug Res.) *5a:* 889–898 (1978).

23   Taeusch, H.W.: Glucocorticoid prophylaxis for respiratory distress syndrome. A review of potential toxicity. J. Pediat. *87:* 616–623 (1975).

24   Wyszogrodski, J.; Taeusch, H.W., and Avery, M.E.: Isoxsuprine-induced alterations of pulmonary pressure/volume relationship in premature rabbits. Am. J. Obstet. Gynecol. *119:* 1107–1111 (1974).

25   Zahn, U.; Zach, H.P. und Sigmund, R.: Über die Möglichkeit der pränatalen Behandlung des Atemnotsyndroms bei Frühgeburten mit Ambroxol. Atemwegs-Lungenkr. *4:* 35–41 (1978).

26   Zimmermann, B. und Merker, H.J.: Über Anwendung der Gewebezucht in Pharmakologie und Toxikologie. Die Möglichkeiten und die Bedeutung von *in vitro*-Methoden zur Substanztestung. Kollogium Reisenburg 1971.

E. Löwenberg, MD, 20th of November Medical Center ISSSTE, México (Mexico)

Prog. Resp. Res., vol. 15, pp. 256–260 (Karger, Basel 1981)

Studies of the Influence of Ambroxol and Dexamethasone on Maturation of Surfactant System in Lungs of Albino Mice[1]

B. Velasquez and J. Sepulveda

Department of Physiology, Pharmacology and Morphological Sciences of the Universidad Autonoma de San Luis Potosi, Medical School, Mexico

Introduction

It has been suggested that Ambroxol stimulates the surfactant system in the fetal lung probably through the increased synthesis and release of intracellular deposits. The mechanism whereby this effect is accomplished is not yet established.

Other authors have reported that in the rat, Bisolvon, a related compound, stimulates the discharge of lamellar bodies and have proposed that this effect is mediated through the action of acid phosphatases that would release surfactant deposits from the lamellar bodies.

On the other hand, it is well known that glucocorticoids given to pregnant animals stimulate lung maturation in all species of mammals studied so far. *Oldenborg* [4] reported that maternal dexamethasone administration induces the appearance of choline phosphotransferase in the lung. We were interested in looking at the histology and the fine structure of the developing fetus, as well as in the newborn, treated either with ambroxol or dexamethasone and to compare them with controls. We studied the last 25% of the total gestation period of the mouse.

Materials and Methods

Pregnant albino CD-1 mice of timed gestational age were used, and starting on the 14th day of pregnancy injections were given up to delivery which occurred at 18.5 or 19 days. After

[1] Partly supported by Böhringer-Ingelheim Mexicana, SA,Conacyt-PNCB 1430, and Coordinacion de la Educacion Superior e Investigacion Cientifica de la SEP'.

delivery no drugs were given but samples were taken from the newborns after 1, 2 and 3 days. The animals received either 30 mg/kg body weight of Ambroxol or dexamethasone 10 mg/kg body weight. The treated animals were compared with two types of controls, i.e., absolute controls where the animals did not receive any treatment at all and saline controls where the animals were given saline solution by intramuscular injection. These animals received a volume proportional to their weight, similar to the one given to the dexamethasone-treated animals. These controls underwent the same stress as those injected with the drugs which could be the source of endogenous corticosteroid release. The total experimental population was formed of 63 pregnant mice and samples were taken from a total of 252 fetuses or newborns, distributed in the following manner: starting on day 14 through delivery, 18 pregnant mice, were given either Ambroxol or dexamethasone, or no drugs as absolute control, and 9 more received saline solution. Either 2 fetuses or 2 newborn animals were taken at random from each litter and samples for light microscopy, electron microscopy and histochemical evaluations using ruthenium red and acid phosphatase were taken.

Results and Comments

No differences between the two types of controls were recorded either at histology or in the fine structures. At the beginning of the study the lung was at the glandular stage. In the treated animals the following differences can be pointed out: The dexamethasone-treated animals appeared to be at a more advanced morphogenetical stage from the 1st experimental day onwards. This was judged by the number of endodermal tubular profiles present in a given field as well as by the dilation and changes in the epithelial height when compared to the others. At the fine structural level there was a higher degree of development of the Golgi areas, associated smooth and coated vesicles as well as multivesicular bodies, all of these being organelles that have been implicated as precursors in the secretion of surfactant. The Ambroxol-treated animals had a closer resemblance to this group than to the controls.

On day 16 a more advanced morphogenesis and differentiation was again observed in the dexamethasone- and now in the Ambroxol-treated animals too. In particular, the Ambroxol-treated animals seemed to have larger glycogen deposits and further development of the Golgi area, associated vesicles and MVB. At this stage, and for the first time, large lamellated structures appeared in the cells near the glycogen deposits in the Ambroxol-treated group. In the dexamethasone-treated animals the degree of morphogenesis and differentiation was higher on day 17. Regarding morphogenesis, the canalicular stage has been initiated and future primitive alveolar cavities can be distinguished from the bronchial tubes. The mesenchyme is diminished with a development of fibroblasts, smooth muscle cells and collagen deposits.

Type one and two cells can be recognized in the epithelial lining. Furthermore, lamellar structures and myelin figures are present in the lumen of these cavities. All these findings suggest an actual developmental acceleration in the dexamethasone-treated animals when compared with the others, where no clearly defined cell types are seen yet. However, the cells also show lamellar bodies and associated structures in the cytoplasm.

On the 18th day the degree of cytodifferentiation is again more advanced in the dexamethasone-treated animals, showing type two pneumonocytes with large accumulations of glycogen; however, in the ambroxol-treated animals there is an apparent increase in the number of cells containing mature and developing lamellar bodies. The presence of these structures even in taller cells that otherwise could not be identified as type two pneumonocytes is remarkable. From the day of delivery and up to the 3rd postnatal day the degree of differences in general development is not very marked among the groups, and the most conspicuous change concerns the appearance of the type two pneumonocytes and the amounts of secreted material found within the primitive alveolar cavities.

Each day, and in a consistent fashion, more osmiophilic lamellar structures and tubular myelin figures were found in the lumens of the ambroxol-treated animals. The type two pneumonocytes have what can be interpreted as a marked secretory activity as judged by a very irregular cell membrane, large lamellar bodies (sometimes seen in the process of exocytosis) as well as abundant MVB and coated vesicles found close to them, indicating the active formation of these organelles. Comparison with the controls or with the dexamethasone-treated animals shows a more regular appearance with smooth cell membranes and smaller and fewer lamellar bodies and less secretion product present in the alveolar lumen.

All these morphological features suggest that although there is an increase in general development in the dexamethasone-treated animals, in the ambroxol-treated animals there appears to be secretory stimulation noticed from the 18th day and up to the 3rd day post partum, as suggested by a marked development of the secretory organelles, e.g., Golgi apparatus, smooth and coated vesicles and MVB, earlier appearance of lamellar bodies and larger amounts of lamellated structures and myelin figures present in the alveolar lumen. This effect appears to be independent of general lung development and should be substantiated with experimental data that could give quantitative information, such as morphometric studies for each group, particularly detecting the appearance and development of the Golgi apparatus, coated vesicles, MVB, and lamellar bodies. Also, studies *in vitro* that would measure the incorporation of radioactive precursors or the amounts recovered from cultures as well as the detection of tensioactive material in both treated and control samples, would give valuable additional information.

References

1   Avery, M.E.: Pharmacologic approaches to the acceleration of fetal lung maturation. Br. med. Bull. *31:* 13–20 (1975)

2   Farrell, P.M. and Avery, M.E.: Hyaline membrane disease. Am. Rev. resp. Dis. *111:* 657–688 (1975).

3   Rooney, S.A., et al.: Studies on pulmonary surfactant. Effects of cortisol administration to fetal rabbit on lung phospholipid content, composition and biosynthesis. Biochim. biophys. Acta *450:* 121–130 (1976).

4   Oldenborg, V. and Van Golde, L.M.G.: Activity of choline phosphotransferase, lysolecithin Acyltransferase in the developing mouse lung. Biochim. biophys. Acta *489:* 454–465 (1977).

5   Renovanz, H.D.: The anti-atelectasis factor considered in relation to bisolvon and its metabolite VIII. Scand. J. resp. Dis. *90:* suppl. 90, pp. 59–64 (1974).

Dr. B. Velasquez, Department of Physiology, Pharmacology and Morphological Sciences of the Universidad Autonoma de San Luis Potosi, Medical School, Mexico (Mexico)

Discussion

Tierney asked whether other cells, for instance Clara cells, were also stimulated and was answered in the affirmative. This was supported by *von Wichert* reporting experiments that showed that Ambroxol raised the phospholipid content in the liver as well as in the lung. He mentioned other experiments showing that Ambroxol restores the PC synthesis in animal experiments with tissue slices in a model with bromcarbomide intoxication but not in a model with peritonitis. Hence, there may be two actions of Ambroxol, one which changes the cellular structure of different lung cells and another which acts on the phospholipid metabolism directly. *Van Golde* discussing these problems said that there is so far no information where this drug acts on phospholipid metabolism, but it is possible that the drug acts on phospho-diesterase and cAMP. The problem, as pointed out by *Bauer,* is that Ambroxol acts as strongly as theophylline on phosphodiesterase but does not reach sufficient plasma levels to also expect such an effect *in vivo*. It may be that such an action is also possible at a much lower dosis.

Prog. Resp. Res., vol. 15, pp. 261–268 (Karger, Basel 1981)

Use of Surfactant to Prevent Respiratory Distress Syndrome[1]

Colin Morley and Alec Bangham[2]

Department of Paediatrics, University of Cambridge and Biophysics Unit, ARC, Babraham, Cambridge

We know from the work of others [1–3] that natural surfactant instilled into the lungs of premature animals aids expansion, prevents alveolar collapse and improves lung compliance and oxygenation, as well as prolonging survival. This encourages us to think that it should be possible to put surfactant into the lungs of surfactant deficient babies either to prevent or to treat RDS. However, it is unlikely that we will be able to use lung wash surfactant for this treatment because: (1) it is difficult to obtain in adequate amounts; (2) it contains proteins to which the baby may become sensitised; (3) it contains bacteria and viruses, and (4) it would be difficult to purify without altering its properties.

Therefore, although this may seem to be the most rational form of treatment, what is needed for bulk production and safety is a simple nontoxic artificial surfactant which can easily be satisfactorily delivered to these very small ill babies. In order to produce an artificial surfactant we need to understand the chemical and physical properties which are most important to its function. We believe the following are the most important properties which any artificial surfactant must contain: (1) the substances must have no acute or chronic toxicity; (2) they must spread as a monolayer rapidly and spontaneously over an air/water interface to produce an equilibrium surface tension of approximately 25 mNm^{-1}; (3) they must contain a reservoir of surfactant molecules from which molecules can spread on to the surface to replace any which may have been lost; (4) under

[1] Supported by Grants from MRC Birthright and the National Fund for Research into Crippling Diseases.
[2] *C.M.* acknowledges the help and encouragement of Prof. *J.A. Davis.*

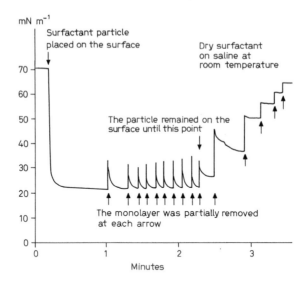

Fig. 1. When a particle of dry surfactant is placed on the surface of water a monolayer spontaneously spreads and the surface tension falls to   25 mNm⁻¹. This monolayer is immediately and repeatedly replenished from the particle as long as the particle remains on the surface to donate molecules.

compression or during expiration the monolayer must not buckle or break but solidify and produce low surface tensions, and (5) they must be in such a form that they can be delivered easily to the lungs of an ill premature baby.

How can these properties be achieved in an artificial surfactant? First, only phospholipids need to be used. Whether protein is a vital part of surfactant *in vivo* is debatable but we can show that all the necessary properties of an artificial surfactant can be produced without protein. This will obviously reduce the toxicity since phospholipids are nontoxic and probably bacteriostatic. Secondly, the phospholipids should be used in a dry and probably particulate form, that is, they must not be used either mixed in water or as a mist in water, because in that case they become liposomes and lose their surface activity. Natural lung surfactant obtained by washings and centrifugation is effective as a replacement surfactant and it is surface active in water. However, it does contain the surfactant as particulate lamellar bodies. Lamellar bodies have recently been shown by *Grathwohl et al.* [4] to contain no free water. There are in fact less than nine

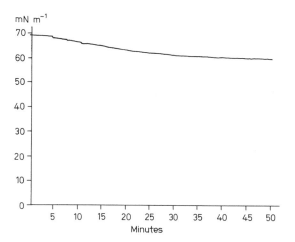

Fig. 2. When the subphase contains surfactant mixed in water and any existing mono-layer has been removed, the wet surfactant is very slow at donating molecules to the surface and lowering surface tension. Sonicated surfactant in the subphase. The original monolayer had been removed. Room temperature.

molecules of water per phospholipid molecule. So lamellar bodies actually function as 'dry' particles of surfactant from within the water. When *Ikegami et al.* [5] sonicated natural surfactant they found it had lost its ability to restore pressure volume curves in surfactant-deficient lungs. This process would have converted the lamellar bodies to liposomes. *Morley and Bangham* [6] have shown that surfactant in a dry particulate form is very effective in spontaneously and rapidly spreading as a monolayer to a surface tension of 25 mNm^{-1} (fig. 1). In this dry state the particles float on the surface and donate more molecules to the surface when required in order to maintain an intact monolayer. When we took the dry surfactant and thoroughly mixed it with the water by sonication, it exhibited very poor surface activity (fig. 2). Molecules slowly reached the surface from the myelin forms under the surface but not at a speed which would be any use physiologically. When we applied this same wet surfactant as a nebulised mist it had very little surface activity (fig. 3). In water, phospholipids form liposomes and the molecules are not easily released from these structures. It therefore seems unlikely that an artificial surfactant thoroughly mixed with water will be as useful as surfactant in a dry particulate form. However, it may be that the phospholipid molecules could also be used

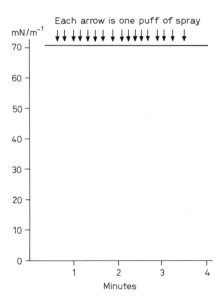

Fig. 3. When hydrated surfactant is nebulised and sprayed onto water as a fine mist at room temperature, few molecules are available to spread on the surface so there is little fall in surface tension.

dispersed in some nonaqueous, nontoxic carrier. This would probably allow the surfactant molecules rapid access to the surface but its use raises problems. First, the molecules of the carrier substance would also enter the surface layer and alter the characteristics of the surfactant film – probably by preventing the development of low surface tensions and solidification under compression, and secondly the carrier substances likely to be used may well have some toxicity and therefore be dangerous in use.

What should be the composition of an artificial surfactant? We know the composition of natural surfactant. It is a complex mixture of substances, mainly lipids, dipalmitoylphosphatidylcholine (DPPC) is the commonest phospholipid and therefore probably plays an important role in the physical properties of surfactant. It has two important features: (1) Being a saturated phospholipid it has a very stable molecular structure with the approximate shape of a cylinder [7]. When these molecules are all packed together tightly in a monolayer their physical shape allows them to form a very rigid structure that will not buckle unless extreme pressures are applied [8]. This is the important feature of surfactant in preventing

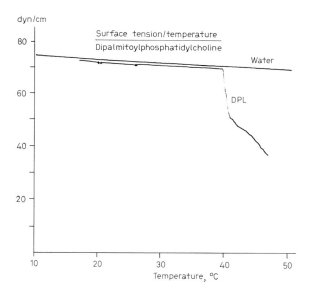

Fig. 4. Pure DPPC is not surface active below 41°C.

atelectasis. (2) DPPC has a transition temperature at 41°C from crystalline to liquid crystal [9]. Below 41°C *pure* DPPC is not spontaneously surface active and on its own it would be a useless surfactant at 37°C (fig. 4).

Any attempt to use pure DPPC alone as an artificial surfactant is unlikely to succeed. Therefore, some other substances must be mixed with DPPC to lower its transition temperature so that it is spontaneously surface active at body temperature. Various surfactant components will do this but they are usually either toxic in large doses, or, like cholesterol, get into the surface monolayer and prevent the development of low surface tensions and solidification under compression. Unsaturated phosphatidyl-glycerol (UPG) seems to be the most promising additive substance to DPPC. It is nontoxic and at body temperature it spreads spontaneously to form a monolayer at 25 mNm^{-1} (fig. 5), but it buckles on compression (fig. 6). When mixed with DPPC in the right proportions the transition temperature is below 37°C so both substances spread together to form a satisfactory monolayer. On compression extremely low surface tensions and a solid film can easily be formed (fig. 7). In consequence it would appear that no other additional substances are needed for an artificial surfactant. This mixture will form a fine powder which if kept dry could be blown into the

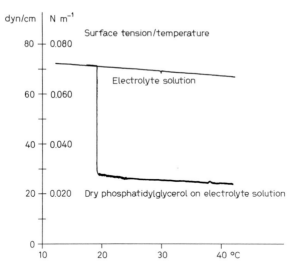

Fig. 5. Unsaturated phosphatidylglycerol spontaneously spreads a monolayer on an aqueous surface at room temperature.

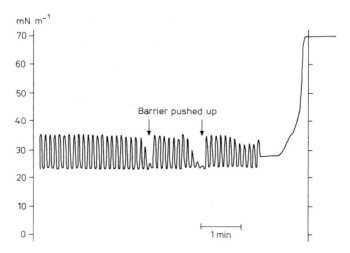

Fig. 6. EGG phosphatidylglycerol at 37°C. When a monolayer of unsaturated phosphatidylglycerol is compressed 40% at 10 cpm, it will not solidify and cannot sustain low surface tensions.

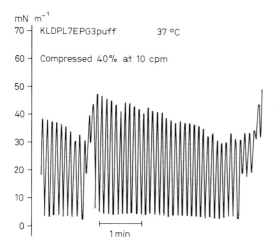

Fig. 7. When a mixture of DPPC and UPG (7:3) is placed on water it spontaneously spreads a monolayer at 37°C. On compression by 40% the surface tension appears to fall and the monolayer solidifies.

lungs down an endotracheal tube. It can be sterile, is nontoxic, and could be produced cheaply and easily in bulk.

What are the problems therefore of treating RDS with artificial surfactant? (1) Bronchiolar lesions and hyaline membranes develop rapidly in surfactant deficient lungs. Once they are formed artificial surfactant is unlikely to be very effective. (2) The lungs of surfactant-deficient premature babies are very wet and any artificial surfactant may prove ineffective in very wet lungs. (3) Premature babies produce small amounts of inefficient surfactant molecules. If these are on the air liquid interface they will hinder spontaneous spreading of and thus may inhibit the effect of the artificial surfactant. (4) Any artificial surfactant has to spread to the alveoli from the trachea and bronchi. It may be prevented from spreading by proteins, mucous or cilia or expelled from the lungs up the trachea by respiratory movements. (5) Respiratory distress in babies may not just be due to surfactant deficiency but to poor muscle tone, inadequate respiratory drive and structurally imature lungs.

This has been a theoretical paper based on our experience but it has drawn on the knowledge from various centres around the world [10, 11] that artificial surfactants can be produced and may be able to ameliorate RDS at least to a limited extent.

Summary

Although natural surfactant can be used to improve lung function in surfactant-deficient states, a nontoxic artificial surfactant is needed. Its desirable physical properties and composition and problems in its use are discussed.

References

1   Enhorning, G. and Robertson, B.: Lung expansion in the premature rabbit fetus after tracheal deposition of surfactant. Pediatrics 50: 58–66 (1972).

2   Nilsson, R.; Grossmann, G., and Robertson, B.: Lung surfactant and the pathogenesis of neonatal bronchiolar lesions induced by artificial ventilation. Pediat. Res. 12: 249–255 (1978).

3   Enhorning, G.; Hill, D.; Sherwood, G.; Cutz, E.; Robertson, B., and Bryan, C.: Improved ventilation of prematurely delivered primates following tracheal deposition of surfactant. Am. J. Obstet. Gynec. 132: 529–536 (1978)

4   Grathwohl, C.; Newman, G.; Phizackerley, P., and Town, M.: Structural studies on lamellated osmiophitic bodies isolated from pig lung. Biochim. biophys. Acta 552: 509–518 (1979).

5   Ikegami, M.; Hesterberg, T.; Nosaki, M., and Adams, F.: Restoration of lung P.V. curves with surfactant. Comparison of nebulisation versus instillation and natural versus synthetic surfactant. Pediat. Res. 11: 178–182 (1977).

6   Morley, C. and Bangham, A.: Physical and physiological properties of dry lung surfactant. Nature, Lond. 271: 162–163 (1978).

7   Shah, D. and Shulman, J.: The ionic structure of lecithin monolayers. J. Lipid Res. 8: 227–233 (1967).

8   Gaines, G.: in Insoluble monolayers at liquid-gas interfaces, pp. 150–180 (Interscience, New York 1966).

9   Albon, N. and Sturtevant, J.: Nature of the gel to liquid crystal transition of synthetic phosphatidylcholines. Proc. natn Acad. Sci. USA 75: 2258–2260 (1978.

10 Ikegami, M.; Silverman, J., and Adams, F.: Restoration of lung pressure. Volume characteristics with various phospholipids. Pediat. Res. 13: 777–780 (1979).

11 Fujiwara, T.; Tanaka, Y., and Takei, T.: Surface properties of artificial surfactant in comparison with natural and synthetic surfactant lipids. IRCS med. Sci. 311: 7 (1979).

C. Morley, MD, Department of Paediatrics, University of Cambridge,
Cambridge CB2 2QQ (England)

Discussion

Answering a question by *Obladen, Morley* said that he had used his preparation on babies but until now he can only say that it did not do any harm to the babies.

Prog. Resp. Res., vol. 15, pp. 269–278 (Karger, Basel 1981)

Treatment of the Premature Rabbit Neonate with Supplementary Surfactant

Bengt Robertson

Department of Pediatric Pathology, Karolinska Institutet, Stockholm

Introduction

Since deficiency of lung surfactant is an essential factor in the pathogenesis of the respiratory distress syndrome (RDS), it would seem logical to prevent, or treat, this disease by administration of supplementary surfactant into the airways. Earlier studies on premature newborn experimental animals have shown that treatment with natural surfactant has indeed a beneficial effect on alveolar expansion and pulmonary pressure-volume characteristics [1, 4, 6–10, 16, 28]. The purpose of our current work is to further define, under *in vivo* conditions, the physical and physiological properties of an optimal lung surfactant, and to develop an experimental model which can be used for testing the efficacy of various synthetic surfactant preparations.

Experimental Model

Premature newborn rabbits are delivered by hysterotomy on day 27 of gestation (full term = 31 days). The animals receive intraperitoneal injections of sodium pentobarbital and pancuronium bromide at birth. Following tracheostomy, they are kept at 37°C in a volume-constant body plethysmograph and subjected to positive-pressure ventilation under standardized conditions. Insufflation pressure and volume changes in the plethysmograph are continuously recorded, and the quasistatic compliance of the lung-thorax system is calculated from registrations of pressure and volume at end-inspiration. For details of the experimental procedure, see *Nilsson et al.* [25].

After a defined period of artificial ventilation, the animals are sacrificed and the lungs are fixed by vascular perfusion for light and electron microscopy. The expansion of the alveolar compartment is evaluated morphometrically in histological sections [6], and the

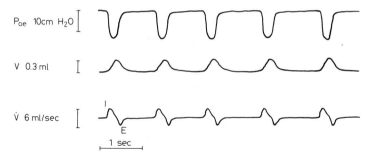

Fig. 1. Registration of lung mechanics during spontaneous ventilation. The tracings were obtained from a surfactant-treated premature newborn rabbit (day 27 of gestation), 60 min after birth. From these tracings the following data were computed: dynamic lung compliance 0.80 ml/cm $H_2O \cdot kg$, inspiratory lung resistance 2.31 cm $H_2O/ml \times sec^{-1}$, expiratory lung resistance 3.16 cm $H_2O/ml \times sec^{-1}$. P_{oe} = oesophageal pressure; V = volume; $\dot{V}$ = flow; I = inspiration; E = expiration.

extent of epithelial lesions in the airways is quantitated according to the protocol of *Nilsson et al.* [25].

In a modification of this experimental model, the animals are tracheostomized without anaesthesia or muscle relaxation. They are ventilated as above, but are at standardized intervals disconnected from the ventilator system and stimulated to breathe spontaneously. Lung mechanics during spontaneous ventilation are measured by means of a fluid-filled oesophageal catheter connected to a pressure transducer, and a specially designed Fleisch tube connecting the tracheal cannula to a differential pressure transducer and an integrator unit. This equipment permits accurate registration of volume changes in the order of 0.005 ml and flow rates $\geqslant$ 0.4 ml/sec, up to a frequency of 8 Hz [21] (fig. 1).

Natural and Artificial Surfactant

A crude but very active preparation of natural surfactant is easily prepared by centrifugation of lung wash from adult experimental animals for 1 h, at 1,000 *g* and 4°C [6]. If the resulting pellet is resuspended in an equal amount of supernatant, the preparation contains approximately 8 mg of phospholipids/ml, 80% of which is lecithin [27].

Various forms of artificial surfactant have been tested with the experimental model described above, in our laboratory and by others. These preparations include dry artificial surfactant [2, 22], emulsified synthetic

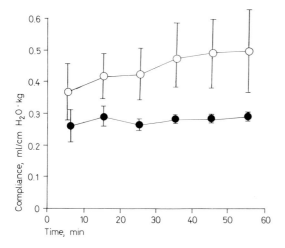

Fig. 2. Lung-thorax compliance ($\bar{X} \pm SD$) at various intervals after birth in surfactant-treated premature newborn rabbits (○) and littermate controls (●) gestational age 27 days. The difference between the groups is statistically significant throughout the period of observation ($p < 0.01$). From *Nilsson et al.* [24].

phospholipids [18, 19], and a mixture of natural and synthetic surfactants produced according to *Fujiwara et al.* [12, 14]. All these preparations have dipalmitoylphosphatidylcholine and unsaturated phosphatidylglycerol as main components.

Observations Following Treatment with Supplementary Surfactant

When a surfactant-deficient premature rabbit neonate (day 27 of gestation) is subjected to intermittent positive pressure ventilation, a comparatively high peak pressure is needed to maintain a tidal volume in the order of 10 ml/kg, and there is no improvement in lung-thorax compliance during the 1st h after birth (fig. 2). In histological lung sections these animals have poor aeration of the alveolar compartment and already within 5–10 min of ventilation, there is necrosis and desquamation of epithelial cells, with early hyaline membrane formation [26] (fig. 3 and 4a).

Tracheal instillation of natural surfactant (50 µl) leads to a significant increase in lung-thorax compliance during artificial ventilation [25] (fig. 2).

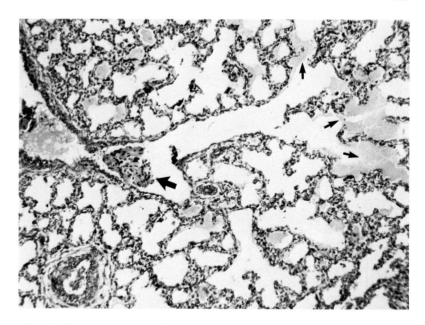

Fig. 3. Epithelial lesions induced by artificial ventilation in a premature newborn rabbit (gestational age 27 days). This animal was ventilated for 1 min with a peak pressure of 35 cm H_2O, then for 60 min with a pressure of 22 cm H_2O. There is oedema and/or unresorbed fetal pulmonary fluid in many terminal airspaces (e.g., small arrows) and prominent necrosis and desquamation of bronchiolar epithelium (big arrow). HE. × 110.

A similar effect has been documented in surfactant-treated animals breathing spontaneously when disconnected from the ventilator system (fig. 1, table I). Treatment with natural surfactant also prevents the development of bronchiolar epithelial lesions during artificial ventilation [25]. This is not simply due to the fact that instillation of surfactant leads to increased lung compliance and that therefore the treated animals can be ventilated with a lower peak pressure. Recent experiments have revealed that treatment with surfactant prevents the development of epithelial lesions even when the animals are ventilated with the same high pressure as required to maintain 'adequate' ventilation in controls (tidal volume ca. 10 ml/kg); under these conditions the compliant lungs of the surfactant-treated animals become greatly overventilated but the epithelium of conducting airways nevertheless remains largely intact (fig. 4).

Improved lung-thorax compliance has also been documented in the

Table I. Dynamic lung compliance during spontaneous ventilation in surfactant-treated premature newborn rabbits and littermate controls (gestational age 27 days). Values were obtained after 30 min of artificial ventilation [23]

Treatment at birth	n	Compliance, ml/cm $H_2O \cdot$ kg	
		mean	range
Natural surfactant	9	0.49	0.06–1.24
Controls	15	0.07	0.01–0.32
$p <$		0.01	

premature rabbit neonate following administration of dry artificial surfactant [23], emulsified synthetic surfactant [20], or a mixture of natural and synthetic surfactants [13] (table II). It is not clear, however, whether dry or emulsified synthetic surfactant preparations prevent the development of epithelial necrosis and hyaline membranes during artificial ventilation. On the other hand, *Fujiwara et al.* [13] have reported that their mixed 'artificial surfactant' is as effective as natural surfactant in preventing epithelial lesions in premature newborn rabbits.

Comments

Natural surfactant has a number of important functional characteristics, which can be summarized as follows:

(1) It has a chain melting temperature of about 28 °C [17] and therefore spreads rapidly on an aqueous hypophase at body temperature. Its equilibrium surface tension at 37 °C is in the order of 27 mN/m [3], but when surface is compressed, surface tension can be reduced to nearly zero [3,5].

(2) It improves the pressure-volume characteristics of lung preparations from surfactant-deficient premature newborn animals and enhances the air expansion of the alveolar compartment [4, 6–10, 16, 28].

(3) As emphasized above, tracheal administration of natural surfactant also improves *in vivo* lung compliance in premature newborn animals and prevents the development of bronchiolar epithelial lesions during artificial ventilation [1,25].

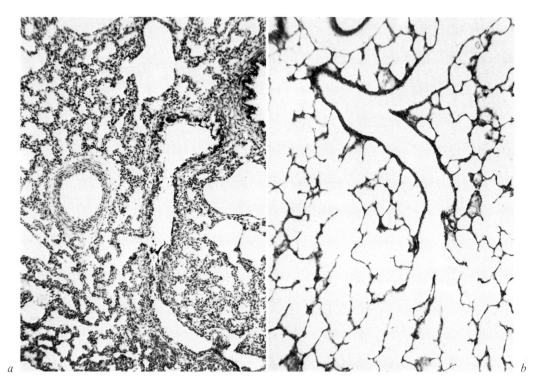

Fig. 4. Lung sections from premature newborn rabbits (gestational age 27 days) ventilated for 10 min with standardized high insufflation pressure (35 cm H$_2$O). HE. × 193. *a.* Control animal showing poor aeration of the terminal airspaces. The bronchiole in the central part of the field shows evidence of epithelial necrosis, with nuclear dust from desquamated cells covering the surface of remaining epithelium. *b.* Animal receiving tracheal instillation of natural surfactant before onset of ventilation. The parenchyma is well aerated and the bronchiolar epithelium is intact.

In our efforts to define a synthetic surfactant preparation with optimal physiological properties, we use physical models (differential thermal analysis, surface balance, pulsating bubble) for initial screening of various substances. Synthetic surfactants with promising *in vitro* properties are then tested on lung preparations from premature newborn animals. Compounds which can improve the pressure-volume characteristics of such lung preparations are further tested under *in vivo* conditions, using the experimental model described above. Although *in vivo* lung compliance can be improved in the premature rabbit neonate by a variety of

Table II. Lung-thorax compliance during artificial ventilation of premature newborn rabbits (gestational age 27 days) after tracheal deposition of various synthetic surfactant preparations. Values are given as $\bar{x} \pm SD$

Type of synthetic surfactant	Duration (min) of artificial ventilation	Compliance, ml/cm H$_2$O · kg		p <	Reference
		experimental animals	controls		
'Dry'	60	0.27 ± 0.38* (n = 12)	0.07 ± 0.06 (n = 17)	0.05	23
Emulsified	10	0.59 ± 0.17 (n = 13)	0.35 ± 0.27 (n = 16)	0.02	17
Mixed natural and synthetic surfactants	60	0.48 ± 0.10 (n = 10)	0.23 ± 0.04 (n = 10)	0.001	13

* Range: 0.02–1.08.

natural and synthetic surfactant preparations, it seems that so far only natural surfactant including the mixture of natural and synthetic surfactants used by *Fujiwara et al.* [13]) offers effective prophylaxis against epithelial lesions during artificial ventilation. Furthermore, our findings indicate that in order to prevent these lesions, supplementary surfactant should be administered as soon as possible and preferably at birth [23].

Necrosis of the bronchiolar epithelium and hyaline membranes are characteristic findings in infants with RDS. Such lesions have been described also in premature infants dying from respiratory insufficiency shortly after birth, even in babies who have not been subjected to artificial ventilation [11, 15]. There is convincing evidence, however, that these bronchiolar lesions tend to become aggravated if the surfactant-deficient infant is treated with artificial ventilation, especially if a high insufflation pressure has to be applied [29]. These clinical observations and the present experimental findings suggest that the artificially ventilated premature rabbit neonate might serve as a useful model of RDS. Any surfactant preparation which improves lung compliance and prevents the development of epithelial lesions in this experimental situation should have a similar effect in infants suffering from surfactant deficiency at birth.

Summary

The artificially ventilated premature rabbit neonate can serve as a useful experimental model of the respiratory distress syndrome. These animals have a low lung compliance and when subjected to artificial ventilation they consistently develop epithelial necrosis and hyaline membranes in terminal conducting airways. Lung compliance is improved and the development of epithelial lesions prevented if the animals are treated with tracheal instillation of natural surfactant before onset of ventilation. A beneficial effect on neonatal lung compliance has also been observed following treatment with dry synthetic surfactant or an emulsified mixture of surfactant lipids. It is not clear, however, whether these synthetic preparations are as efficient as natural surfactant in preventing epithelial lesions during artificial ventilation.

Acknowledgements

The data presented above were obtained jointly with Dr. *Roland Nilsson* and Dr. *Gertie Grossmann* (Department of Pediatric Pathology, Karolinska Institutet, Stockholm, Sweden), Dr. *Burkhard Lachmann* (Research Institute for Lung Diseases, Berlin-Buch, GDR) and Dr. *Colin Morley* (Department of Child Health, Medical School, Cambridge, UK).

Financial support was provided by the Swedish Medical Research Council (Project No. 3351), The Swedish National Association against Heart and Chest Diseases, Allmänna BB:s Minnesfond, Expressens Prenatalforskningsfond, and Karolinska Institutets fonder.

References

1 Adams, F.H.; Towers, B.; Osher, A.B.; Ikegami, M.; Fujiwara, T., and Nozaki, M.: Effects of tracheal instillation of natural surfactant in premature lambs. I. Clinical and autopsy findings. Pediat. Res. *12*: 841–848 (1978).

2 Bangham, A.D.; Morley, C.J., and Phillips, M.C.: The physical properties of an effective lung surfactant. Biochim. biophys. Acta *573*: 552–556 (1979).

3 Clements, J.A.: Functions of the alveolar lining. Am. Rev. resp. Dis. *115* (suppl., June), pp. 67–71 (1977).

4 Enhörning, G.: Photography of peripheral pulmonary airway expansion as effected by surfactant. J. appl. Physiol.; Resp. Environ. Exercise Physiol. *42*: 976–979 (1977).

5 Enhörning, G.: Pulsating bubble technique for evaluating pulmonary surfactant. J. appl. Physiol.; Resp. Environ. Exercise Physiol. *43*: 198–203 (1977).

6 Enhörning, G.; Grossmann, G., and Robertson, B.: Tracheal deposition of surfactant before the first breath. Am. Rev. resp. Dis. *107*: 921–927 (1973).

7 Enhörning, G.; Grossmann, G., and Robertson, B.: Pharyngeal deposition of surfactant in the premature rabbit fetus. Biol. Neonate *22*: 126–132 (1973).

8 Enhörning, G.; Hill, D.; Sherwood, G.; Cutz, E.; Robertson, B., and Bryan, C.: Improved ventilation of prematurely-delivered primates following tracheal deposition of surfactant. Am. J. Obstet. Gynec. *132*: 529–536 (1978).

9   Enhörning, G. and Robertson, B.: Lung expansion in the premature rabbit fetus after tracheal deposition of surfactant. Pediatrics, Springfield 50: 58–66 (1972).

10 Enhörning, G.; Robertson, B.; Milne, E., and Wagner, R.: Radiologic evaluation of the premature newborn rabbit after pharyngeal deposition of surfactant. Am. J. Obstet. Gynec. 121: 475–480 (1975).

11 Finlay-Jones, J.M.; Papadimitriou, J.M., and Barter, R.A.: Pulmonary hyaline membrane: light and electron microscopic study of the early stage. J. Path. 112: 117–124 (1974).

12 Fujiwara, T.; Maeta, H.; Chida, S., and Morita, T.: Improved pulmonary pressure-volume characteristics in premature newborn rabbits after tracheal instillation of artificial surfactant. IRCS med. Sci. 7: 312 (1979).

13 Fujiwara, T.; Maeta, H.; Chida, S., and Morita, T.: Improved lung-thorax compliance and prevention of neonatal pulmonary lesion in prematurely delivered rabbit neonates subjected to IPPV after tracheal instillation of artificial surfactant. IRCS med. Sci. 7: 313 (1979).

14 Fujiwara, T.; Tanaka, Y., and Takei, T.: Surface properties of artificial surfactant in comparison with natural and synthetic surfactant lipids. IRCS med. Sci. 7: 311 (1979).

15 Gandy, G.; Jacobson, W., and Gairdner, D.: Hyaline membrane disease. I. Cellular changes. Archs. Dis. Childh. 45: 289–310 (1970).

16 Grossmann, G.: Expansion pattern of terminal air-spaces in the premature rabbit lung after tracheal deposition of surfactant. Pflügers Arch. 367: 205–209 (1977).

17 Grossmann, G.; Larsson, I.; Nilsson, R.; Robertson, B.; Rydhag, L., and Stenius, P.: Synthetic surfactant as substitute for pulmonary surfactant; in Georgiev, Lung lipid metabolism, mechanisms of its regulation, and alveolar surfactant. Proceedings of the second international symposium, Varna, May 20–24, 1979 (Publishing House of the Bulgarian Academy of Sciences, Sofia, in press).

18 Grossmann, G. and Larsson, I.: Surface properties of natural surfactant in comparison with various synthetic lipids. IRCS med. Sci. 6: 478 (1978).

19 Grossmann, G. and Larsson, I.: Improved pulmonary pressure/volume characteristics in premature newborn rabbits after tracheal instillation of synthetic surfactants. IRCS med. Sci. 6: 479 (1978).

20 Grossmann, G.; Larsson, I.; Nilsson, R.; Robertson, B.; Rydhag, L., and Stenius, P.: Emulsified synthetic surfactant; surface properties and effect on neonatal lung expansion during artificial ventilation. Abstract. Path. Res. Pract. 165: 100 (1979).

21 Lachmann, B.; Grossmann, G.; Nilsson, R., and Robertson, B.: Lung mechanics during spontaneous ventilation in premature and fullterm rabbit neonates. Respir. Physiol. 38: 283–302 (1979).

22 Morley, C.J.; Bangham, A.D.; Johnson, P.; Thorburn, G.D., and Jenkin, G.: Physical and physiological properties of dry lung surfactant. Nature, Lond. 271: 162–163 (1978).

23 Morley, C.; Robertson, B.; Lachmann, B.; Nilsson, R.; Bangham, A.; Grossmann, G., and Miller, N.: Artificial surfactant and natural surfactant. Comparative study of the effects on premature rabbit lungs. Archs. Dis. Childh. (in press).

24 Nilsson, R.; Grossmann, G., and Robertson, B.: Pathogenesis of neonatal bronchiolar epithelial lesions induced by artificial ventilation. IRCS med. Sci. 5: 272 (1977).

25 Nilsson, R.; Grossmann, G., and Robertson, B.: Lung surfactant and the pathogenesis

of neonatal bronchiolar lesions induced by artificial ventilation. Pediat. Res. *12:* 249–255 (1978).

26   Nilsson, R.; Grossmann, G., and Robertson, B.: Bronchiolar epithelial lesions induced in the premature rabbit neonate by short periods of artificial ventilation. Acta path. microbiol. scand. [A] (in press).

27   Robertson, B.: Neonatal pulmonary mechanics and morphology after experimental therapeutic regimens; in Scarpelli and Cosmi, Reviews in perinatal medicine, vol. 4 (Raven Press, New York, in press).

28   Robertson, B. and Enhörning, G.: The alveolar lining of the premature newborn rabbit after pharyngeal deposition of surfactant. Lab. Invest. *31:* 54–59 (1974).

29   Taghizadeh, A. and Reynolds, E.O.: Pathogenesis of bronchopulmonary dysplasia following hyaline membrane disease. Am. J. Path. *82:* 241–264 (1976).

B. Robertson, MD, Department of Pediatric Pathology, Karolinska Institutet; Pediatric Pathology Research Laboratory, St. Görans Hospital, S–112 81 Stockholm (Sweden)

Discussion

Dr. *Enhörning* asked why natural surfactant prevents these lesions. Dr. *Robertson* answered that treatment with surfactant will facilitate uniform aeration of the pulmonary parenchyma, thereby reducing longitudinal shear forces which in the surfactant-deficient lung might cause disruption of the Gronchiolar epithelium.

Prog. Resp. Res., vol. 15, pp. 279–284 (Karger, Basel 1981)

Results of Animal Experiments in the Therapy of RDS-Syndrome

W. Endell, R. Grosspietzsch, F. Klink and L.v. Klitzing

Frauenklinik der Medizinischen Hochschule, Lübeck

Introduction

One of the principle problems in premature delivery is the absence or relative insufficiency of lung surfactant leading to respiratory distress syndrome (RDS) of the newborn infant. According to a general agreement the chemically defined lecithin DPPC (dipalmitoylphosphatidylcholine) is the greatest single fraction of the surface-active material [5]. This surfactant factor is synthesized in the alveolar cells type II and stored in the osmiophilic lamellar bodies [1].

Today, glucocorticoids are used for the prophylaxis and therapy of hyaline membrane disease with the aim of stimulating the phospholipid synthesis. Since the administration of glucocorticoids in cases of imminent prematurity is not without risk and not in every case practicable [8], we suggest the administration of the necessary lecithin to the fetus as a better choice.

Results

In animal experiments with rabbits and mini-pigs, a placental barrier of DPPC of an intravenous injection to the pregnant animal was verified. As the experimental animal, we chose the Göttinger mini-pig. As an omnivora it resembles man not only in its metabolism but also in chemistry of its lipids in the serum and organs, especially in the distribution of the fatty acids [4, 9].

Generally, the gestation of mini-pigs lasts 114 days. Approximately, the 95th day of gestation corresponds to prematurity in man with

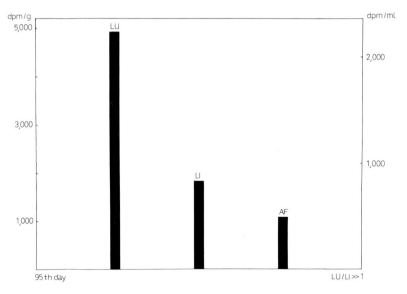

Fig. 1. Comparison of radioactivity in fetal organs – lung (LU), liver (LI) and amniotic fluid (AF) – in mini-pigs on the 95th day of gestation.

imminent RDS. Since we found out during our first studies that a total placental barrier exists for lecithin [6], for the following experiments we chose to apply lecithin directly into the amniotic fluid. As test substance we used ^{14}C-labelled lecithin.

The animal experiments were carried out under halothane anesthesia. The animals were laparotomized and a single dosis of 10^6 dpm ^{14}C-lecithin was injected into each amniotic sack. After injection we sewed up the abdominal cavity and stopped the anesthesia [2]. 6 h after application the animals were relaparotomized and all the fetuses were delivered alive under simultaneous withdrawal of the amniotic fluid. After 15 min of spontaneous respiration we processed the fetal organs: lungs, liver, intestines and brain. The analysis included thin-layer chromatography separation and subsequent autoradiography, described by *Grosspietzsch et al.* [3].

Figure 1 demonstrates the amount of radioactivity in the tissue homogenates of lung, liver and amniotic fluid in animals on the 95th day of gestation. A comparison of the columns shows that selectively more lecithin was assimilated in the lung tissue.

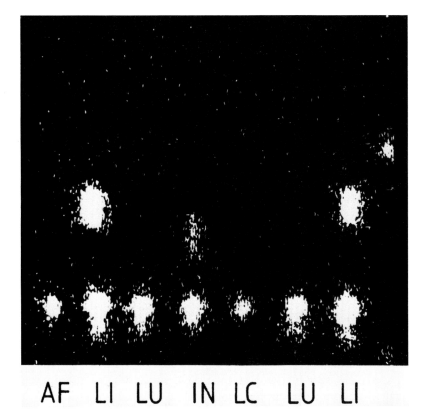

AF LI LU IN LC LU LI

Fig. 2. Thin-layer chromatographic plate of fetal organ homogenates – lung (LU), liver (LI), intestines (IN), amniotic fluid (AF) and standard lecithin (LC) – in mini-pigs on the 95th day of gestation.

The ^{14}C-radioactivity of lecithin and lecithin metabolites on the thin-layer chromatography plate is shown in figure 2. The spots of the lung tissue and the amniotic fluid can be seen at the same height as the lecithin standard. Additionally, another metabolite can be demonstrated in the liver tissue.

Figure 3 represents the amount of radioactivity in the fetal organ homogenates of lung, liver and amniotic fluid. This is the same application of lecithin as before, but on the 110th day of gestation, that means only a few days anterpartum. Contrary to the animals with 95 days of gestation, we found a higher amount of lecithin in the liver compared with the lung

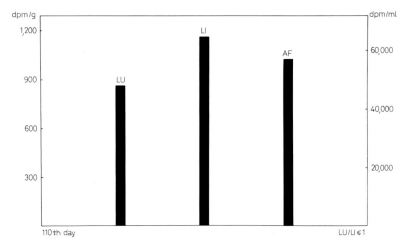

Fig. 3. Comparison of radioactivity in fetal organs – lung (LU), liver (LI) and amniotic fluid (AF) – in mini-pigs on the 110th day of gestation.

tissue. The radioactivity of lecithin is again demonstrated on the thin-layer chromatography plate (fig. 4). The spot of the lecithin found in the lung tissue is to be seen on the same level as the standard lecithin. Lavage of the fetal alveolar surface was performed with a wash solution which has a lipophilic and a hydrophilic part. After a defined duration of the lavage ^{14}C-lecithin could also be measured in the lavage fluid.

Discussion

These results of our animal experiments convinced us that intra-amnially applied lecithin was absorbed in the intestinal tract [7], followed by distribution into the whole fetal organism. Around the 95th day of gestation, we could demonstrate that an especially high assimilation rate is found in the fetal lungs. In the last phase of gestation we found that a lower lecithin assimilation in the fetal lungs takes place if the fetal organism is completely matured.

On the basis of these convincing results we performed similar experiments in humans as we regard this treatment as an effective prophylaxis of the RDS syndrome.

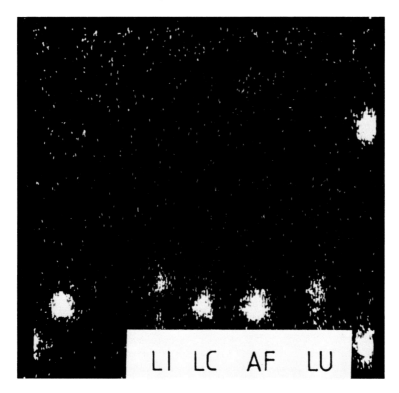

Fig. 4. Thin-layer chromatographic plate of fetal organ homogenates – lung (LU), liver (LI), amniotic fluid (AF) and standard lecithin (LC) – in mini-pigs on the 110th day of gestation.

Summary

In case of deficiency of the lecithin DPPC (dipalmitoylphosphatidylcholine) in newborn infants, the clinical picture of a respiratory distress syndrome will develop. In our experiments we studied whether an assimilation of this lecithin on the alveolar surface can be obtained by intra-amnial application of DPPC. These experiments were performed in rabbits and Göttinger mini-pigs. After intra-amnial application we could in all experiments demonstrate the presence of lecithin (DPPC) or its metabolites in the fetal lung tissue homogenate and on the fetal alveolar surface. The detection of ^{14}C-labelled lecithin was performed by measurement of radioactivity and thin-layer chromatography in the organ homogenates. Labelled lecithin was also demonstrated in the lavage fluid of the alveolar surface. We may conclude that intra-amnially applied DPPC attains the fetal lungs and acts efficiently postpartum.

References

1. Campiche, P.M.; Gautier, A.; Hernandez, E.I., and Reymond, A.: An electron microscope study of the fetal development of human lung. Pediatrics, Springfield *32:* 976 (1963).

2. Endell, W.; Grosspietzsch, R.; Klink, F.; Holtz, W. und v. Klitzing, L.: Experimentelles Modell zur intraamnialen Applikation von Substanzen bei multiparen Versuchstieren. Res. exp. Med. *172:* 155–159 (1978).

3. Grosspietzsch, R.; Klink, F.; v. Klitzing, L.; Endell, W. und Oberheuser, F.: Intraamniale Applikation von ^{14}C-markiertem Lecithin beim Menschen 13 Stunden ante partum. Geburtsh. Frauenheilk. *37:* 527–531 (1977).

4. Insull, W.: Fatty acid composition of human adipose tissue related to age, sex and race. Am. J. clin. Nutr. *20:* 13–23 (1967).

5. King, R.J. and Clements, J.A.: Surface active materials from dog lung. Method of isolation. Am. J. Physiol. *223: 707–714 (1972).*

6. v. Klitzing, L.; Endell, W.; Grosspietzsch, R.; Klink, F. und Oberheuser, F.: Prophylaxe und Therapie des Atemnot-Syndroms bei Frühgeborenen mit Lezithin. Fortschr. Med. *95:* 501–504 (1977).

7. Lekim, D. und Graf, E.: Tierexperimentelle Studien zur Pharmakokinetik der «essentiellen» Phospholipide (EPL). Drug Res. *26:* 1772–1782 (1976).

8. Ohrlander, S.A.; Gennser, G.M., and Grennert, L.: Impact of betamethasone load given to pregnant women on endocrine balance of fetoplacental unit. Am. J. Obstet. Gynec. *123:* 228–236 (1975).

9. Zöllner, N. and Tacconi, M.: Über die Konzentration einiger Lipide in Serum und Organen des Miniaturschweins und ihre Abhängigkeit vom Alter des Tieres. Z. ges. exp. Med. *145:* 326–334 (1968).

Dr. med. vet. W. Endell, Frauenklinik of the Medical Faculty, Ratzeburger Allee 160, D-2400 Lübeck (FRG)

Prog. Resp. Res., vol. 15, pp. 285–292 (Karger, Basel 1981)

Enrichment of Surfactant in Preterm Fetal Lungs by Intra-Amnial Administration of DPL

R. Grosspietzsch, A. Fenner, F. Klink and L. v. Klitzing

Departments of Gynecology and Obstetrics and Neonatology, School of Medicine, Lübeck

Introduction

Based on animal experiments [1, 5–7] we were able to administer intra-amnially ^{14}C-labelled DPL (DPPC) in three particular clinical cases in the 18th, 28th and 38th weeks of gestation to fetuses not expected to live because of sanctionated interruptio and anencephalus [2–4]. The intra-amnial injection of phosphatidyl-(^{14}C-methyl)-choline (Amersham-Buchler) dissolved in a 10% Intralipid® emulison (1 part Intralipid + 9 parts NaCl, total radioactivity 1.28×10^8 dpm) was performed between 12 and 14 h ante partum. In all three cases an analysis of organ homogenates of fetal lung and liver was carried out just as an additionally lung lavage shortly postmortem. Figures 1 and 2 demonstrate the results of an anencephalus in the 38th week of gestation by thin-layer chromatography (modified Kynast/Saling method) and autoradiography (beta-camera, Berthold-Friesecke).

These investigations revealed that DPL, administered intra-amnially ante partum, is swallowed by the fetus, absorbed in the intestines, metabolized during liver passage and appears as the resynthesized lecithin molecule on the alveolar surface with the reservation that the influence of fetal respiratory movements could not be taken into consideration.

Clinical Methods

In all cases where preterm delivery seemed unpreventable even though high dosage intravenous tocolysis was performed, intra-amnial administration of DPL was carried out on the occasion of the first amniocentesis for collecting amniotic fluid for surfactant determi-

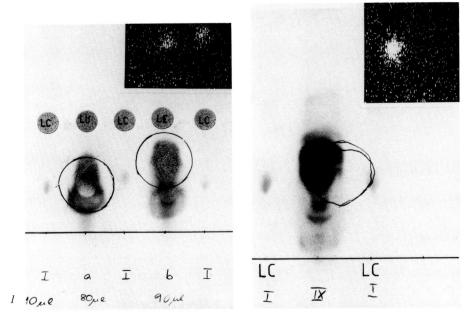

Fig. 1. Thin-layer chromatography of the fetal organs, LU (lung) and LE (liver), after intra-amnial administration of DPL in an anencephalic fetus of the 38th week of gestation. LC = Lecithin standards. Top right: Autoradiography of the same preparation.

Fig. 2. Thin-layer chromatography of the lavage fluid of the same fetus and treatment as above (fig. 1). IX = Lavage fluid; I = Lecithin standard. Top right: Autoradiography of the same preparation.

nation. As immature we considered all fetuses up to the 36th week of gestation at most, normally up to the 35th week. Figure 3 gives an example of the clinical procedure in these cases. After the intra-amnial administration of DPL we tried to delay delivery as long as possible by intravenous tocolysis for at least 12 or 14 h.

All postpartum investigations of the immature newborns were done separately by the Department of Neonatology. The investigator in this department did not know if DPL was administered or not in order to guarantee an unprejudiced diagnosis.

Results and Discussion

The probands were grouped according to the following criteria (table I): DPL received or not, and RDS during the first 72 h of life or not.

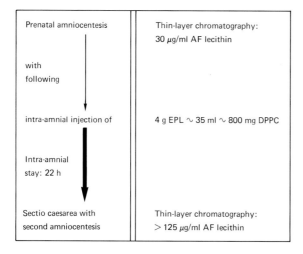

Fig. 3. Scheme of clinical procedure of intra-amnial administration of DPL (DPPC) in a case of a 24-year-old primipara, 36th week of gestation, EPH-gestosis (RR 230/130).

Table I. Proband material.

			n
Total number			20
Group I	DPPC received,	no RDS,	11
Group II	no DPPC received,	no RDS,	2
Group III	DPPC received,	RDS,	3
Group IV	DPPC received, severe	RDS,	0
Group V	no DPPC received,	RDS,	4
Group VI	DPPC received, tocolysis successful, delivery beyond 37th week of gestation		13

Grouping was done according to (1) DPL received or not; and (2) RDS suffered or not. Group VI comprises those 13 infants in whom tocolytic treatment given simultaneously with intra-amnial DPL was successful. Thus, these infants remained *in utero* until term. They were all well at birth.

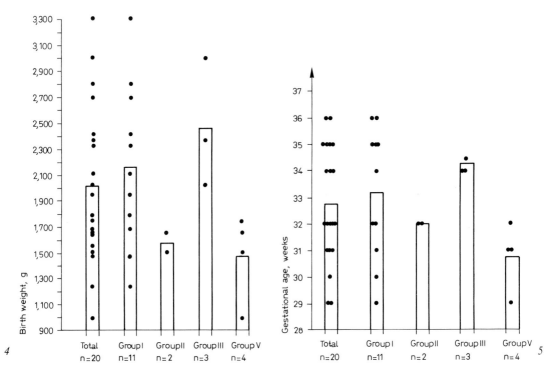

4

5

Fig. 4. Birth weight of probands; means and individual values. The groups are too small for comparison.

Fig. 5. Gestational age of probands; means and individual values.

Group IV is a hypothetical one since there were no severe cases of RDS among the infants who had received DPL into their amniotic liquid. Group VI comprises those infants in whom tocolytic therapy, attempted in all cases, was successful, pregnancy continued and all 13 infants were in good health when they were born.

With respect to birth weight (fig. 4), gestational age (fig. 5) and 1-min Apgar scores (fig. 6) no systematic group differences are apparent. Thus, homogeneity of proband material may be assumed even though the numbers are still small.

Prenatally determined concentration of lecithin in amniotic liquid showed a wide variety of values ranging from 12.1 to 62.5 µg/ml (fig. 7).

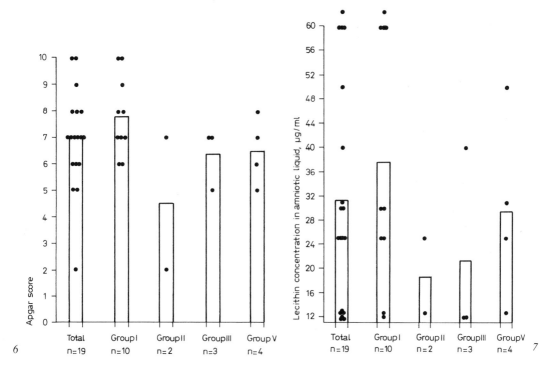

Fig. 6. Apgar scores (1 min) of probands; means and individual values. Note that there are no Apgar scores below 6 in group I (DPL received, no RDS).

Fig. 7. Lecithin concentration in amniotic liquid; means and individual values. The numbers are too small for comparisons. Note that of the 10 figures given for group I, 6 are in the subnormal and 4 in the normal range.

In group I (DPL received, no RDS) the mean value is 37.71 µg/ml, although 6 of the 10 tests performed, yielded results of values at or below 30 µg/ml.

PCO_2 values (fig. 8) over the first 48 h of life show a similar pattern in all groups: initially high PCO_2 with a rapidly falling tendency. In those infants suffering from RDS (groups III and V), The PCO_2 remained somewhat higher. Needless to say, these patients received respiratory assistance, otherwise their results would look much more pathologic. A near mirror image of the PCO_2 values is seen in the courses of the pH

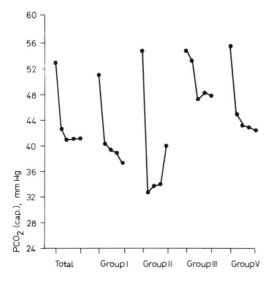

Fig. 8. Course of PCO_2 tension over the first 48 h (initial, 6 h, 12 h, 24 h, 48 h) of extra-uterine life (means). Rapid improvement occurs during the first 6–12 h (group III). Improvement is less complete in groups III and V in which all infants were suffering from RDS. The values would probably be even higher if the patients had not received ventilatory assistance.

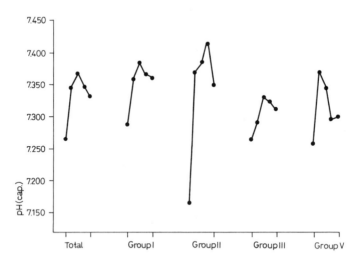

Fig. 9. Course of pH values over the first 48 h of extrauterine life (means). Expectedly, the values represent a nearly complete mirror of the PCO_2 tension shown in figure 5.

Table II. Statistical analysis (chi square test) of results: summary of cases

	RDS	No RDS	Σ
DPPC	3	11	14
No DPPC	4	2	6
Σ	7	13	20

The numbers are too small to reach the level of statistical significance.

$\chi^2 n = 2.05$ (NS).

results (fig. 9). Again, RDS patients (groups III and V) tend to level off at a lower pH value.

A summary of the preliminary results is given in table II: of our 20 cases, there were 7 with and 13 without RDS. Of the 13 no-RDS infants, 11 had received intra-anmiotic DPL. These results do not reach the level of statistical significance at this point. Of particular interest is the fact that three infants did suffer from RDS although DPL had been administered intraamnially. All cases were mild, one of them in an infant of a diabetic mother.

In conclusion it may be stated that DPL, administered intraamnially in cases of threatening premature delivery, apparently has no harmful effects. Of the 14 cases treated with DPL in our study, 11 had no RDS and 3 a mild form. It is hoped that future cases will follow the same pattern. The study is continuing.

Summary

Based on animal and clinical investigations intra-amnial administration of DPL (DPPC) in cases of unpreventable preterm delivery was carried out in 33 patients for prophylaxis of RDS. In 13 cases tocolytic therapy was successful, pregnancy continued and all 13 infants were in good health when born. In 20 cases premature delivery was unpreventable. In 14 of these 20 cases DPL could be injected intra-amnially early enough. Of these 20 cases there were 7 with and 13 without RDS. Of the 13 no-RDS infants 11 had received intra-amniotic DPL. 3 infants did suffer from RDS although DPL had been administered intra-amnially. Ale these 3 cases were very mild, one of them in an infant of a diabetic mother.

References

1   Endell, W.; Grosspietzsch, R.; Klink, F.; Holtz, W. and Klitzing, L.v.: Experimentelles
    Modell zur intraamnialen Applikation von Substanzen bei multiparen Versuchstieren.
    Res. exp. Med. *172:* 155 (1978).

2   Grosspietzsch, R.; Klink, F.; Klitzing, L.v.; Oberheuser, F. und Endell, W.: Fallstudie
    zur Plazentagängigkeit von ^{14}C-markiertem Lecithin. Fortschr. Med. *94:* 1465 (1976).

3   Grosspietzsch, R.; Klink, F.; Klitzing, L.v.; Endell, W. und Oberheuser, F.: Intraam-
    niale Applikation von ^{14}C-markiertem Lecithin beim Menschen 13 Stunden ante
    partum. Geburtsh. Frauenheilk. *37:* 527 (1977).

4   Klink, F.; Klitzing, L.v.; Endell, W.; Grosspietzsch, R. und Oberheuser, F.: Verlaufsbe-
    obachtung nach intraamnialer Lezithin-Applikation bei vorzeitiger Schnittentbindung.
    Fortschr. Med. *94:* 1783 (1976).

5   Klitzing, L.v.; Endell, W.; Grosspietzsch, R.; Klink, F. und Oberheuser, F.: Prophylaxe
    und Therapie des Atemnotsyndroms bei Frühgeborenen mit Lezithin. Fortschr. Med.
    95: 501 (1977).

6   Oberheuser, F.; Endell, W.; Grosspietzsch, R.; Klink, F. und Klitzing, L.v.: Intraam-
    niale Applikation von Lecithin bei hochträchtigen Kaninchen. Z. Geburtsh. Perinat.
    181: 59 (1977).

7   Okoh, O.; Grosspietzsch, R. und Klitzing, L.v.: Hat die Ernährungsgewohnheit mit
    hohem Anteil an Palmöl (Palmitinsäure) Einfluss auf die niedrige Atemnotsyndromrate
    in Nigeria? Mschr. Kinderheilk. (in press).

R. Grosspietzsch, MD, Department of Gynecology and Obstetrics, School of Medicine,
Ratzeburger Allee 160, D-2400 Lübeck (FRG)

Discussion

Morley asked if there was an increase in the amount of effective material in the lung and
Grosspietzsch answered that the reason for the experiments was to look if the labels go down
to the alveoli and not to measure the amount of lecithin within the alveoli. *Bauer Biberach*
felt from the reported paper that the material would be heavily metabolized rather than go
down to the alveoli unchanged and *Grosspietzsch* confirmed that the same distribution of the
label would be achieved by using palmitic acid as tracer.

Prog. Resp. Res., vol. 15, pp. 293–296 (Karger, Basel 1981)

Prophylaxis of RDS of the Newborn with Steroids

J. Hary and W. Rindt

Department of Gynecology and Obstetrics, University of Saarland, Homburg

Though the treatment of premature labor is highly effective, the problem of premature deliveries with consecutive RDS is still important. Consequent therapy of late abortion often results in premature delivery and, in some cases, premature delivery is unavoidable by medical reasons. Thus, the prevention of RDS is one of the most important challenges in perinatal medicine [9].

In experimental animals, as well as in humans, several drugs have been investigated as to whether they are able to increase surfactant in the amniotic fluid and/or decrease the incidence of RDS when given to the mother.

Epinephrine, the β-adrenergic substance isoxsuprine (3) and carnitine were effective in experimental animals, as well as thyroxine, TRH and prolactin. The bromhexine metabolite Ambroxol is obviously useful in RDS prevention in the human (1, 5, 8), whereas tocolytic substances apparently failed to do so. 17-β-Estradiol proved to be effective in the rabbit (4), whereas conjugated estrogens showed no significant effect in humans (6). It seems to be possible to improve prevention of RDS with estradiol valerianate. Among the corticosteroids the following compounds have been investigated: hydrocortisone, dexamethasone, betamethasone and prednylidene (1, 2, 6, 8).

Which of the corticosteroids mentioned is to be chosen favorably is difficult to say. Clinical data do not allow any advantages between these substances to be established. Whichever is chosen, we must pay attention to its pharmacokinetic parameters such as half-life of elimination and biologic potency. Here some examples of therapeutic schemes:

Table I. Increase in L/S ratio in % of initial values

Week	Prednylidene				Ambroxol			
	single	values		mean	single	values		mean
26	2			–	–			–
27	–			–	4			–
28	0			–	0	9		–
29	0	32		–	38	25	3	22
30	41	17	33	32	57	23	26	32
31	29	36	23	29	63	13	20	32
32	41	6	35	28	19	27		23
33	21	23	16	20	17	46	27	30
34	15	33	18	22	34	23		26

An increase in L/S ratio is evident. Simultaneously, the incidence of RDS is relatively low.

Daily dosage: 2 × 12 mg betamethasone, 2 × 8 mg b-disodium phosphate, + 6 mg b-acetate; 6 mg b-disodium phosphate, + 6 mg b-acetate (Celestan®). Amniocentesis, 12 mg dexamethasone, (3 days); amniocentesis, 3 days 60 mg prednylidene i. v. (Decortilen®).

Further modifications are certainly possible and even necessary in several difficult cases. The total amount of steroids within these schemes is influenced by the clinical follow-up, the effectiveness of tocolytic therapy, if necessary, and by evaluation of the L/S ratio or similar parameters, if technically possible. One fact should be mentioned explicitly: the interval between drug application and delivery. In order to obtain an appropriate reaction, one should strive for a 24- to 48-hour interval.

Our own experience extends over 5 years. Primarily, prevention was performed with betamethasone and later on with prednylidene (16-methylene-prednisolone; Decortilen®) in a dosage as mentioned in the scheme given above.

Our results date from a comparative study with prednylidene and ambroxol. They represent our general experience in the prevention of RDS with corticosteroids (table I, II).

Altogether, we believe, in accordance with most of the authors working in this field, that the application of corticosteroids to the mother is, after the 28th week of gestation, a very useful prevention of RDS.

Table II. Incidence of respiratory distress syndrome

	Prednylidene week of gestation			Ambroxol week of gestation		
	26–32	33–37	38	26–32	33–37	38
Newborns	1	11	–	–	12	9
RDS	1	2	–	–	2	–
Incidence	1	0.18	–	–	0.17	–

Nevertheless, a final proof, based on a double-blind trial with a placebo group is not possible. On the other hand, we cannot conceal that the degree of improvement is not really satisfactory. *Schutte et al.* [7] recently published a study which could not prove a positive effect of corticosteroids compared to placebo. It certainly must be respected that all of his patients were treated with orciprenaline infusions.

This opens the field for discussion as to whether it is necessary to combine different drugs or not. As it is a highly interesting problem, further investigations should be carried out, for example, on a combination of corticosteroids with Ambroxol. Such a combination could perhaps reduce the total amount of corticosteroids. There is no doubt that such a reduction should be achieved, since there are several absolute contraindications such as tuberculosis, diabetes, inflammatory diseases like amnion infection, intestinal ulcus and severe EPH syndrome.

In case of reduction of the corticosteroid dosage for RDS prevention, at least some of these cases could benefit and give birth to a mature and healthy child.

References

1   Becker, H.H.: RDS-Prophylaxe mit Ambroxol und Prednilyden; Diss. Homburg/Saar (1979).
2   Caspi, E.; Schreyer, P.; Weinraub, Z.; Bukoysky, I., and Tamir, I.: Changes in amniotic fluid L/S ratio following maternal dexamethasone administration. Am. J. Obstet. Gynec. *122*: 327–31 (1975).
3   Enhörning, G.; Chamberlain, D.; Contreras, C.; Burgoyne, R., and Robertson, B.: Isox-suprine infusion to the pregnant rabbit and its effect on fetal lung surfactant. Biol. Neonate *35*: 43–51 (1979).

4   Khosla, S.S. and Rooney, S.A.: Stimulation of fetal lung surfactant production by administration of 17-β-estradiol to the maternal rabbit. Am. J. Obstet. Gynec. *133:* 213–216 (1979).

5   Muller-Tyl, E.: Clinical trial thomae (Biberach/Riss, 1977).

6   Spellacy, W.N.; Buhi, W.C.; Rigall, F.C., and Holsinger, K.L.: Human amniotic fluid L/S ratio changes with estrogen and glucocorticoid treatment. Am. J. Obstet. Gynec. *115:* 216–18 (1973).

7   Schutte M.F.; Treffers, P.E.; Koppe, J.G.; Breur, W., and Fildet Kok, J.C.: Klinische toepassing van corticosteroiden ter bevordering van de foetale long-rijpheid. Ned T. Geneesk. *123:* 420–427 (1979).

8   Schwenzel, W., und Jung, H.: Pränatale Behandlungsmethoden zur Vermeidung eines ANS bei frühgeborenen Kindern. Gynäkologe *8:* 198–205 (1975).

9   Worthington, D.; Maloney, A.H.A., and Smith, B.T.: Fetal lung maturity. Mode of onset of premature labor. Obstet. Gynec. *49:* 275–279 (1977).

Dr. Hary., J., Department of Gynecology and Obstetrics, University of Saarland, D-6650 Homburg/Saar (FRG)

Discussion

In answer to a question by *Morley* it was said that it is impossible to compare the incidence of RDS in a treated population against an untreated control group.

Prog. Resp. Res., vol. 15, pp. 297–300 (Karger, Basel 1981)

Combined Glucocorticoid-Fenoterol Treatment in the Prevention of the Respiratory Distress Syndrome

Analysis of Beneficial and Adverse Effects

B. Liedtke and S. Schneider

Abteilung Gynäkologie und Geburtshilfe, Medizinische Fakultät, Rheinisch-Westfälische Technische Hochschule (Vorstand: Prof. *H. Jung*), Aachen

Introduction

One of the most severe complications in premature newborns is the development of the respiratory distress syndrome (RDS) due to lung immaturity. RDS is responsible for the majority of neonatal deaths in premature infants.

Materials and Methods

For the prevention of RDS in neonates we treated women in premature labor with glucocorticoids, namely betamethason (2 injections with an interval of 24 h, each injection containing 1.5 ml of Celestan Depot®) or methylprednisolone (3 injections with intervals of 24 h, each injection containing 60 mg of Decortilen®). Fenoterol was the only uterine relaxant used in all cases. We report about 483° women treated in the above-mentioned way. The distribution of age in our patients shows a peak between 22 and 29 years. The youngest patient was 14 and the oldest 43 years old. In our collective there are significantly more primiparous than multiparous women. The distribution in our patients is in accordance with other investigators.

The incidence of RDS in the treated collective is compared with untreated controls at the same gestational age. There was no possibility for corticoid treatment in this control group because of sudden parturition due to failure of tocolysis, or because of contraindications against corticoid therapy as gestosis or severe diabetes mellitus.

To compare the incidence of RDS we formed four groups of different gestational ages: first group 28–30 weeks of gestation, second group 31–34 weeks of gestation, third group 35–37 weeks of gestation and fourth neonat es over 38 weeks of gestation. In accordance with the literature the last group was excluded, because an RDS due to lung immaturity after 38 weeks of gestation cannot be expected.

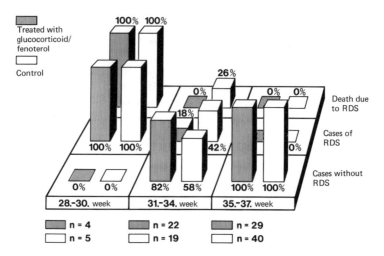

Fig. 1. Incidence of RDS in the treated collective compared with untreated controls at the same gestational age. A statistically significant reduction in the incidence of RDS is apparent in the corticoid group with deliveries between 31 and 34 weeks of gestation.

Results

In prematures who were less than 30 weeks we could not find an effect of corticoid therapy in the prevention of RDS. All treated and untreated prematures developed RDS and did not survive. There was no case of RDS in the neonates born in or after the 35th week of gestation, neither in the treated group nor in the control group

A statistically significant reduction in the incidence of RDS is apparent in the corticoid group with deliveries between 31 and 34 weeks of gestation (fig. 1). In the untreated group we found 42% RDS. Only 18% of the treated prematures developed RDS. There were mild and transitory distresses, but there was no neonatal death by RDS in the treatment group. In the untreated controls 26% of severe RDS responsible for neonatal deaths were found.

In a previous study of our group, we reported on the danger of adverse effects of the combined corticoid and fenoterol therapy. Therefore, we analyzed the treated group of 483 women for adverse effects. At first we investigated the levels of electrolytes in blood serum. We found no indication of retention of sodium, but in 31% mildly lowered levels of sodium in the serum likely due to gestation (table I).

Table I. Analysis of adverse effects of the combined corticoid and fenoterol therapy

	Cases	n	%
Levels of electrolytes in blood serum			
Sodium: no indication of retention	–	483	0
Postassium: level <3.7 mEq/l	144	483	29.8
level <3.3 mEq/l	45	483	9.3
Signs of maternal infection			
Elevated body temperature up to 37.9 °C	34	483	7.0
Fever over 38 °C	7	483	1.4
Infection of the amnion	2	483	0.4
Aggravation of the symptoms of gestosis	9	483	1.9

Analyzing the serum potassium levels during therapy and on the first 2 days after therapy, levels were below 3.7 mEq/l in 144 cases (29.8%) and below 3.3 mEq/l in 45 cases (9.3%). During the following days there was a trend for normalization. 4 days after glucocorticoid therapy, but with continously applied fenoterol, we found no cases with potassium levels lower than 3.3 mEq/l.

During glucocorticoid therapy we gave ampicillin to each patient. Looking for signs of maternal infection we found 34 cases (7.8%) with elevated body temperature up to a maximum of 37.9°C. 7 patients (2%) developed fever with temperature more than 38°C. 2 cases had signs of an infection of the amnion. In these months we do a prospective study to determine, if ampicillin has a protective effect against infection during the described glucocorticoid therapy.

Careful attention should be payed to patients with preexisting symptoms of gestosis. In our study 9 patients (1.9%) suffered an aggravation of their gestosis. This is why women with signs of severe gestosis should be excluded from this therapy.

The examination for adverse effects in the neonates revealed only 12 cases (2.4%) with signs of infection of uncertain etiology.

Summary and Discussion

Our results clearly indicate that corticoid treatment is capable of influencing fetal lung maturity. The clinical effect could be statistically confirmed for the prematures who had been born between 31 and 34 weeks of gestation. In our collective neonates born after 35 weeks of gestation did not develop RDS neither in the treatment group nor in the control group.

Analyzing the adverse effects, we emphasize the danger of an aggravation of gestosis caused by the combined corticoid-fenoterol therapy and a mild risk of infection. A dangerous adverse effect of the combined corticoid-fenoterol therapy is the lowered serum potassium level. This disturbance of serum electrolytes is caused by the betamimetic drug fenoterol at the onset of therapy. This effect is slightly aggravated by the combined corticoid therapy. This electrolyte imbalance caused by therapy might worsen preexistent cardiac disease. Therefore, an intensive clinical observation is necessary. Under the mentioned conditions, we see an indication for combined glucocorticoid-fenoterol therapy in patients with premature labor until the 36th week of gestation.

References

1 Halberstadt, E. und Gerner, R.: Die biochemische Organreifung des Feten und ihre Induktion. Z. Geburtsh. Perinat. *183:* 1–11 (1979).
2 Hallmann, M. and Gluck L.: Development of the fetal lung. J. perinatal Med. *5:* 3–31 (1977).
3 Liedtke, B. und Jung, H.: Gefahren der Kombinationsbehandlung von Beta-Mimetika mit Kortikoiden zur fetalen Reifebehandlung; in Jung und Friedrich, Fenoterol (Partusisten®) bei der Behandlung in der Geburtshilfe und Perinatologie, p. 215 (Thieme, Stuttgart 1978).
4 Liggins, G.C. and Kitterman, J.A.: Pharmacology of fetal lung maturation; in Thalhammer, Baumgarten and Pollak, Perinatal medicine, Sixth European Congress, Vienna, p. 19 (Thieme, Stuttgart 1979).

Dr. med. B. Liedtke, Abteilung Gynäkologie und Geburtshilfe, Medizinische Fakultät, RWTH Aachen, Goethestrasse 27–29, D-5100 Aachen (FRG)

Discussion

Dr. Harvy asked about different corticoids, but using betamethason or methylprednisolone there are no differences; patients with severe gestosis are excluded from the treatment. Administration of cortisol in patients with ruptured membranes does not lead to an increase in infections.

Prog. Resp. Res., vol. 15, p. 301 (Karger, Basel 1981)

Experimental Results with Steroids in Adult Animals

D.S. Tierney

Manuscript not submitted.

Discussion

In answering a question by *Wolf,* Tierney confirmed that the lungs of the oxygen-treated animals were severely damaged which is observed more often in older than in younger rats. Speculating on a question by *van Golde* about the reasons for the different utilization of lyso-phosphatidylcholine in the slices experiments and the infused-lung experiments, *Tierney* answered that the different tissue preparation (the perfused lung experiments are done in relatively intact lungs) may be a result of the release of enzymes by tissue damage produced by tissue slicing. Answering a question by *von Wichert,* Tierney discussed the problems of relating the values to DNA, wet weight or dry weight, and he pointed out that the best way might be to relate the metabolic data to whole lung.

Prog. Resp. Res., vol. 15, pp. 302–307 (Karger, Basel 1981)

Content of Phospholipid and Dipalmitoyl Lecithin in the Fetal Rat Lung after Treatment with Steroids, Thyroxine and Ambroxol[1]

A. Lohninger, P. Krieglsteiner, W. Riedl, F. Fischbach and G. Blümel

Forschungsinstitut für Traumatologie der AUVA, Wien

There is evidence of glucocorticoids accelerating the development of fetal lungs. This was first published by *Liggins et al.* [3]. In view of the risk which might possibly be involved with steroid treatment, other ways for the prevention of RDS are also being discussed. In the present studies, the effects of betamethasone, thyroxine and Ambroxol on the content of phospholipid and dipalmitoyl lecithin [DPL] in the fetal rat lungs were evaluated.

In the first series of preliminary experiments, betamethasone and Ambroxol was administered to 45 gravid Wistar rats for a period of 4 days prior to premature delivery and a saline solution was given to the controls. These investigations showed only a slight influence of betamethasone and Ambroxol on the content of phospholipid and lecithin in the fetal rat lungs.

With the same experimental procedure, the effects of treatment with betamethasone und thyroxine were investigated in a second series of preliminary experiments by determining the content of phospholipid and DPL on the 18th day of gestation and then compared with a group of controls. When compared with the controls, the groups treated with betamethasone and/or thyroxine showed a slight increase in DPL content.

[1] This paper was supported by a grant from the Allianz-Versicherungs AG.

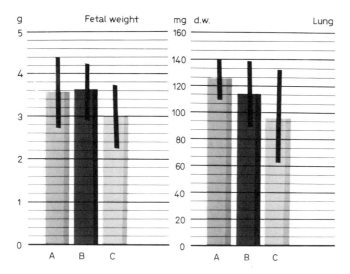

Fig. 1. Fetal weight and weight of the lungs of one litter. A = NaCl; B = T_4; C = betamethasone.

Method

In a third series of experiments, 40 gravid rats having an average weight of 300 g and an expected gestation period of 23 days were treated with NaCl (controls), betamethasone (0.1 mg/kg body weight), and T_4 (150 µg/kg body weight) from the 17th to the 20th day of gestation. The section was carried out on the 21st day of gestation which corresponds to approximately 90% of the total gestation period. After perfusion with a saline solution, the fetal lungs were removed by applying microsurgical methods. All lungs of one litter were combined and processed together so as to preclude any differences which might be due to a different position of the fetuses in the bicorn uterus. Lipid was extracted according to the method of *Folch* [2]. Phospholipids were determined according to the method of *Bartlett* [1].

After separation by TLC, the composition of the lecithin fatty acids was determined in a gas-liquid chromatographical process. DPL was determined by gas-liquid chromatography with glass capillary columns by taking dimyristoyl lecithin as the internal standard [4].

Results

Treatment with betamethasone results in a remarkable decrease of the fetal weight as well as the weight of the fetal lungs, when compared with the control group and the group treated with thyroxine (fig. 1). Neither by

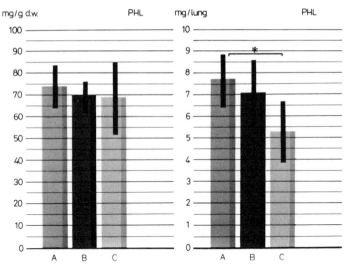

Fig. 2. Total phospholipid (PHL) content in mg/g dry weight of the fetal lung tissue and total PHL content of the lungs of one litter. A = NaCl; B = T_4; C = betamethasone.*p < 0.05.

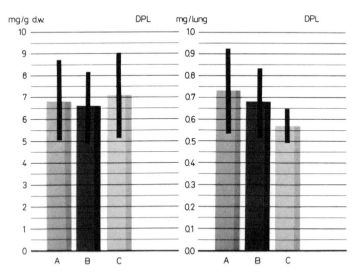

Fig. 3. Dipalmitoyl lecithin (DPL) content in mg/g dry weight of the fetal lung tissue and DPL content in mg of the lungs of one litter. A = NaCl; B = T_4; C = betamethasone.

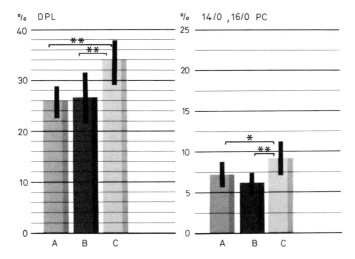

Fig. 4. Percentage of dipalmitoyl lecithin (DPL) on the lecithin species and the percentage of myristoyl-palmitoyl lecithin (14/0, 16/0 PC) on the lecithin species of the fetal lungs. A = NaCl; B = T_4; C = betamethasone. *p < 0.05; **p < 0.01.

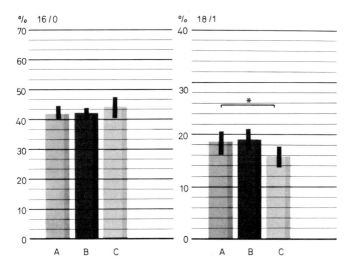

Fig. 5. Percentage of the palmitic acid (16/0) and oleic acid (18/1) on the lecithin fatty acids. A = NaCl; B = T_4; C = betamethasone. *p < 0.05

betamethasone nor by thyroxine could the content of phospholipid in the fetal lungs be increased with regard to that of the controls (fig. 2). The group treated with betamethasone showed a slightly higher content of DPL per gram dry weight than the control and thyroxine group (fig. 3).

After treatment with betamethasone, the percentage of dipalmitoyl lecithin in the other lecithin species was significantly increased as compared with the controls and the other test group (fig. 4). With regard to the T4 group and the controls, the percentage of myristoyl-palmitoyl lecithin in the other lecithin species was also significantly increased in the group treated with betamethasone (fig. 4). After treatment with betamethasone, the fatty acid pattern of the lecithin showed an increase in the palmitic acid and a remarkable decrease in the oleic acid (fig. 5).

Discussion

In accordance with the results of the present paper, *Wu et al.* [5] reported that there was no evidence of the maturation of fetal lungs being accelerated when gravid dams were treated with thyroxine. When thyroxine was either injected into the fetuses directly or into the amniotic cavity, this group showed a higher surface activity of the fetal lungs than the control group.

It seems to be established that glucocorticoids have a stimulating effect on the activity of two enzymes in the CDP-choline pathway in fetal lungs. The results of the present paper show that compared with the control group, a treatment with betamethasone leads to a slight increase in the content of DPL per gram dry weight of the fetal lungs.

As opposed to studies dealing with the incorporation of labelled precursors in the lecithin via the CDP-choline pathway, the DPL determination by way of direct capillary gas chromatography showed a specific effect of betamethasone on the content of DPL in the fetal lung tissue. Since mainly unsaturated lecithin species and only a small portion of the DPL are formed via *de novo* lecithin synthesis, the marked increase in the percentage of DPL in the lecithin species leads to the assumption that steroids also increase the DPL synthesis via the deacyclic and reacyclic cycle.

References

1   Bartlett, G.R.: Phosphorus assay in column chromatography. J. biol. Chem. *234:* 466–468 (1959).

2   Folch, J.: A simple method for the isolation and purification of the total lipids from animal tissues. J. biol. Chem. *226:* 497 (1957).

3   Liggins, G.C. and Howie, R.N.: Controlled trial of anterpartum glucocorticoid treatment for prevention of the respiratory distress syndrome in premature infants. Pediatrics *50/4:* 515 (1972)

4   Lohninger, A. and Nikiforov, A.: Quantitative determination of natural dipalmitoyl lecithin with dimyristoyl lecithin as internal standard by capillary gas-liquid chromatography. J. Chromat. *192:* 185 (1980).

5   Wu, B.; Kikkawa, Y.; Orzalesi, M.M.; Moto Yama, E.K.; Kaibara, M.; Zigas, G.J., and Cook, C.D.: The effect of thyroxine on the maturation of fetal rabbit lungs. Biol. Neonate *22:* 161–168 (1973).

A. Lohninger, MD, Forschungsinstitut für Traumatologie der Allgemeinen Unfall-Versicherungs-Anstalt, A-1200 Wien (Austria)

Prog. Resp. Res., vol. 15, pp. 308–316 (Karger, Basel 1981)

Experimental Results in the Prevention of RDS with α-Tocopherol

H. Wolf, W. Seeger, N. Suttorp and H. Neuhof

Centre of Internal Medicine, Department of Experimental Medicine and Clinical Pathophysiology, Justus-Liebig-University, Giessen

Introduction

Deterioration of surfactant activity is a constant symptom early in shock lung development. The nature of impairment in function and availability of surfactant molecules is not yet clear. Oxidising agents as e.g. ozone, NO_2, long-time exposure to high tension O_2, or in a more indirect way, paraquat intoxication, cause quite similar defects in surface tension activity [9, 27]. In animal experiments pathophysiological and histological findings characteristic of the respiratory distress syndrome (RDS) are constantly induced by microembolisation of pulmonary circulation. In our model [21] we found hypoxic hyperventilation, massive increase in pulmonary vascular resistance with subsequent systemic hypocirculation. Histology shows additional microthrombi formed by aggregated thrombocytes and leucocytes. perivascular and interstitial oedema and atelectatic areas. Surface tension parameters show severe deterioration. The influence of α-tocopherol in high doses as biologic antioxidant on surface tension deterioration is the question of this investigation.

Material and Methods

Model. Awake rabbits of both sexes, weighing 2.13 ± 0.34 kg, breathing ambient room air, were intermittently microembolised via a PVC-catheter placed in the right atrium, by means of aggregated human albumin particles, such as are used for perfusion scintigrams (Tecepart®, Behringwerke). 86% of the particles had diameters between 10 and 40 μm. During the first 3 h 13.5 mg/kg body weight were given (1 mg = $712,000 \pm 157,000$ particles). From

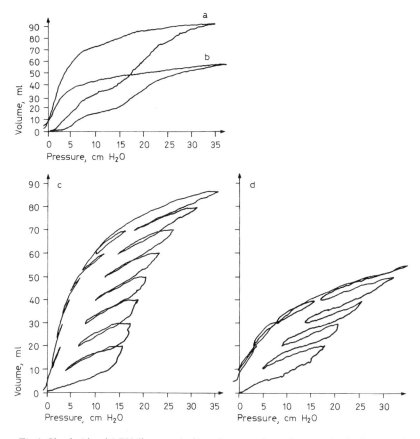

Fig.1. Classic (simple) PV diagrams (a, b) and pneumoloop diagrams (c, d) of a control animal (a, c) and of a microembolised reference animal (b, d).

hour 6 to 8, 12 to 14, and 18 to 20 doses of 4 mg/kg were added. Surviving animals were sacrificed after 24 h.

Pressure-Volume (PV) Diagrams. They were drawn from immediately *post mortem* isolated lungs as classic (simple) PV diagrams and pneumoloop diagrams (fig. 1).

Parameters. Parameters referred to are: (1) $V_{max/K}$ (ml/kg) = Inflated air volume per kilogramme body weight in classic PV diagrams at a constant maximum pressure of 35 cm H_2O. (2) P_K (cm H_2O) = Pressure at the bending point of the deflation curve. The bisector of the angle formed by the tangents in the flattest and steepest part of the deflation curve in classic PV diagrams intersects this curve in the bending point. (3) $C_{Q/pv}$ = Compliance quotient of PV

Table I. Means, standard error, and statistics (one-way analysis of variance) of surface tension parameter

	$V_{max/K}$		P_K		$C_{Q/pV}$		$C_{Q/pl}$		$C_{defl/K}$		$C_{infl/K}$	
	$\bar{x}$	$s_{\bar{x}}$	$\bar{x}$	$s_{\bar{x}}$	$\bar{x}$	$s_{\bar{x}}$	$\bar{x}$	$s_{\bar{x}}$	$\bar{x}$	$s_{\bar{x}}$	$\bar{x}$	$s_{\bar{x}}$
A	44.00	2.40	8.98	0.30	48.13	3.45	5.77	0.41	4.34	0.32	0.74	0.049
B	29.60	1.65	7.30	0.20	26.55	2.37	3.14	0.28	2.07	0.22	0.67	0.033
C	35.08	2.54	6.91	0.31	43.27	3.66	6.43	0.41	3.84	0.32	0.60	0.049
D	31.71	2.40	7.54	0.30	28.20	3.45	3.86	0.41	2.27	0.34	0.60	0.052

	$V_{max/K}$	P_K	$C_{Q/pV}$	$C_{Q/pl}$	$C_{defl/K}$	$C_{infl/K}$
A:B	$p < 0.001$	$p < 0.001$	$p < 0.001$	$p < 0.001$	$p < 0.001$	N.S.
B:C	N.S.	N.S.	$p < 0.001$	$p < 0.001$	$p < 0.001$	N.S.
C:D	N.S.	N.S.	$p < 0.01$	$p < 0.001$	$p < 0.001$	N.S.
B:D	N.S.	N.S.	N.S.	N.S.	N.S.	N.S.

N.S. = Not significant. Group A: n = 9; B: n = 19; C: n = 8; D: n = 9.

diagrams, i.e. the compliance of the steepest part divided by the compliance of the flattest part of the deflation curve. (4) $C_{Q/pl}$ = Compliance quotient of pneumoloops, i.e. the quotient of the compliances of corresponding pneumoloops in deflation and inflation part at $\frac{1}{3}$ maximum volume.

Animal Groups. A (n = 9): Control animals having no microembolisation, but equal amounts of Ringer's solution. B (n = 19): Reference animals being microembolised in the mentioned way. C (n = 8): Microembolised animals which received $dl\alpha$-tocopherol by duodenal tube in a dosage of 50 mg/kg dissolved in 5 ml vegetable oil at 18 h and again 1 h prior to the beginning of microembolisation. D (n = 9): Microembolised reference animals which received only vegetable oil in the same manner without additional tocopherol.

Results

Microembolised animals (table I) which received vegetable oil only (D), do not show any difference to the reference animals (B), whereas the tocopherol group (C) differs significantly to both reference groups (B and D): In the *pneumoloop diagrams* (fig. 2) $C_{Q/pl}$ as well as the compliance of the deflation part ($C_{defl/K}$) show significantly higher levels in the tocopherol-treated animals. On the other hand, the compliance of the inflation

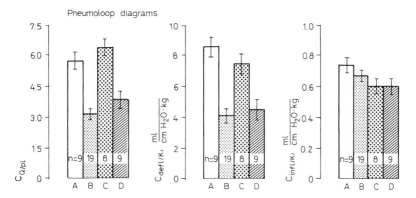

Fig. 2. Surface tension parameters of the pneumoloop diagrams: $C_{Q/pl}$ = Compliance quotient of pneumoloops, $C_{defl/K}$ = compliance of pneumoloops in the deflation part and $C_{infl/K}$ = compliance of pneumoloops in the inflation part, both at ⅓ maximum volume and corrected according to body weight. A = Control (no microembolisation); B = microembolisation; C = microembolisation after tocopherol in vegetable oil; D = microembolisation after vegetable oil only.

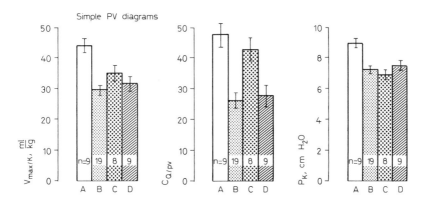

Fig. 3. Surface tension parameters of classic simple PV diagrams: $V_{max/K}$ = Inflated air volume at maximum pressure of 35 cm H_2O per kilogramme body weight; $C_{Q/pv}$ = compliance quotient of PV diagrams; P_K = pressure at the bending point of the deflation curve. For groups A–D see legend 2.

part ($C_{infl/K}$) does not differ significantly. Thus, hysteresis is significantly improved in the tocopherol group. In the *classic PV diagrams* (fig. 3) $C_{Q/pv}$ of the tocopherol group exceeds the values of both reference groups highly significantly, reaching those of the non-microembolised control animals. Taking into consideration the even lower pressure at the bending point (P_K)

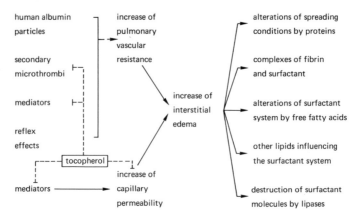

Fig. 4. Interstitial oedema caused by alterations in the pulmonary vascular bed and its influence on the conditions of surfactant activity. Possible tocopherol effects.

and the partially improved $V_{max/K}$ these findings imply the conclusion that tocopherol did not influence the slight depletion of surfactant material, but prevented the predominantly functional impairment of the surfactant system in the microembolisation model.

Discussion

To explain the demonstrated effects of tocopherol on surfactant activity one has to refer to its known antioxidant [23] and membrane-stabilising properties [4, 8, 23]. We see two main pathways in possible peroxidative defects in shock lung pathology, being influenced by tocopherol in the sense of (a) *primary reduction of interstitial oedema* (fig. 4). Interstitial oedema alters the consistence of surfactant hypophase by extravasated proteins, free fatty acids and other lipids [19, 28]. Thus, optimal spreading conditions for surfactant molecules are seriously impaired. Complexes of surfactant with fibrin are possible [13] as well as a destruction of surfactant molecules e.g. by lipases [5]. Tocopherol could diminish interstitial oedema by stabilising the membranes of capillary endothelium [25], lysosomal membranes, and by the reduction of thrombocyte aggregation [16a,b]. The second main pathway is primarily concerned with (b) *surfactant activity amelioration* (fig. 5). It is true that about 75% of the

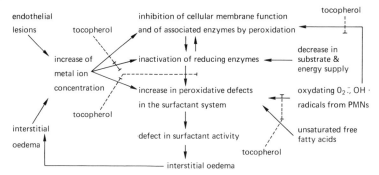

Fig. 5. Functional deterioration of the surfactant system by peroxidative mechanisms, induced by a rise in metal ion concentration, lack of energy supply, oxygen radicals, and unsaturated free fatty acids. Possible tocopherol effects. PMNs = Polymorphonuclear neutrophilic leucocytes.

phospholipids in the surfactant system consist of saturated fatty acids, but the remaining fraction of unsaturated fatty acids is related mostly to the phosphatidyl glycerols, which seem to have a clue function in surfactant activity, as has been indicated by the papers of *Morley* and of *Obladen* (this symposium). So peroxidation of unsaturated fatty acid substituents may impede the spreading conditions of the whole system seriously.

Increased peroxidation in the surfactant system in RDS is possible pursuing at least the following four pathways: (1) Under physiological conditions hydroperoxides are constantly formed in lung lipids [6, 23] and are reduced by NADP-H consuming enzymes as e.g. the glutathion-peroxidase/reductase system [4, 12]. Microembolisation and microthrombosis of pulmonary vessels cause inactivation of enzyme activities by blocking substrate and energy supply. (2) Being easily peroxidised, increased levels of free fatty acids in lung interstitium during shock development cause an augmentation of peroxidative changes in the surfactant system [3, 7, 15]. (3) Leucocytes are able to release various oxygen radicals from their surface [14, 20, 26]. The superoxide dismutase/catalase system seems to be involved in detoxication of these radicals [18] and is known as an extremely fast acting enzyme, but still little is known about the properties of this enzyme under *in vivo* conditions. Thus, leucocytes sticking in the pulmonary vessels [1, 17] may be a cause of increased peroxidation in the surfactant system and in pneumocytal and endothelial cell membranes. (4) Heavy metal ions, especially those linked to proteins, powerfully initiate and catalyse peroxidation [6, 10]. In that way endothelial lesions and, in the sense of a vitious circle, interstitial oedema are able to initiate and to increase peroxidative alterations [6, 10, 23, 28].

In all these pathways tocopherol is presumably protective to the surfactant system by interruption of free radical chain reactions. The physiological effect of tocopherol can be multiplied by the administration of pharmacological doses [2, 11, 22, 24]. As a consequence we give tocopherol clinically (about 30 mg/kg/day) to all patients having developed or are threatened by RDS or are ventilated for a presumably longer period with high inspiratory oxygen tension. We feel that treatment of these critically ill patients has become more successful since we apply the additional tocopherol therapy. However, demonstrable evidence of the effect will be very difficult to achieve in a group of patients as inhomogenous and polypragmatically treated as RDS patients. So we have to count upon further investigations in pulmonary peroxidation pathology as well as we need further experience with antioxidants in RDS therapy.

Summary

Alterations of pulmonary surface tension induced by microembolisation of pulmonary vessels in rabbits are demonstrated, using pressure-volume diagrams of *post mortem* isolated lungs. Dl-α-tocopherol given in pharmacological doses perorally prior to microembolisation prevents functional deterioration of surfactant activity, but does not seem to influence slight depletion of surfactant material. Pathways of pulmonary peroxidation pathology are discussed.

References

1   Bergofsky, E.H.: Pulmonary insufficiency after nonthoracic trauma: shock lung. Am. J. med. Sci. *264:* 93 (1972).
2   Bonetti, E.; Abondanza, A.; Dello Corte, E.D.; Novello, F., and Stirpe, F.: Studies on the formation of lipid peroxides and on some enzymic activities in the liver of vitamin E-deficient rats. J. Nutr. *105:* 364 (1975).
3   Brücke, P.: Die Pathophysiologie der Lungen bei experimenteller Fettembolie; in Lungenveränderungen bei Langzeitbeatmung, p. 282 Wiemers und Scholler, (Thieme, Stuttgart 1973).
4   Chow, C.K.; Reddy, K., and Tappel, A.: Effect of dietary vitamin E on the activities of the glutathione peroxidase system in rat tissues. J. Nutr. *103:* 618 (1973).

5   Cooper, N.; Matsuura, Y.; Murner, E.S., and Lee, W.H.: Analysis of pathologic mechanisms in pulmonary surfactant destruction following thermal burn. Am. Surg. *33:* 882 (1967).

6   Demopoulos, H.B.: The basis of free radical pathology. Fed. Proc. *32:* 1859 (1973).

7   Derbs, C.M. and Jacobvitz-Derbs, D.: Embolic pneumopathy induced by oleic acid. Am. J. Path. *87:* 134 (1977).

8   Diplock, A.T.: Possible stabilizing effect of vitamin E on microsomal, membrane-bound, selenide-containing proteins and drug metabolizing enzyme systems. Am. J. clin. Nutr. *17:* 995 (1974).

9   Dowell, A.R.; Kilburn, K.H., and Pratt, P.C.: Short-term exposure to nitrogen dioxide. Archs intern. Med. *128:* 74 (1971).

10 Haugaard, N.: Cellular mechanisms of oxygen toxicity. Physiol. Rev. *48:* 311 (1968).

11 Kayden, H.J.; Chow, C.K., and Bjornson, L.K.: Spectrophotometric method for determination of tocopherol in red blood cells. J. Lipid Res. *14:* 533–540 (1973).

12 Khandwala, A.; Bernard, J., and Gee, L.: Linoleic acid hydroperoxide: impaired bacterial uptake by alveolar macrophages, a mechanism of oxidant lung injury. Science *182:* 1364 (1973).

13 Lasch, H.G.: Hämostase und Schocklunge. Verh. dt. Ges. inn. Med. *81:* 462 (1975).

14 Mandell, G.L.: Catalase, superoxide dismutase, and virulence of *staphylococcus aureus.* J. clin. Invest. *55:* 561 (1975).

15 Matsuura, Y.; Najib, A., and Lee, W.H.: Pulmonary compliance and surfactant activity in thermal burn. Surg. Forum *17:* 86 (1966).

16a Moncada, S.; Needleman, P.; Bunting, S., and Vane, J.R.: Prostaglandin endoperoxide and thromboxane generating systems and their selective inhibition. Prostaglandins *12:* 323–329 (1976).

16b Gryglewsi, R.J.; Bunting, S.; Moncada, S.; Flower, R.J., and Vane, J.R.: Arterial walls are protected against deposition of platelet thrombi by a substance (prostaglandin X) which they make from prostaglandin endoperoxides. Prostaglandins *12:* 685–713 (1976).

17 Ratliff, N.B.; Wilson, J.W.; Hackel, D.B., and Martin, A.M.: The lung in hemorrhagic shock. II. Observations on alveolar and vascular ultrastructure. Am. J. Path. *58:* 353 (1970).

18 Rister, M. and Baehner, R.L.: The alteration of superoxide dismutase, catalase, glutathione-peroxidase, and NAD(P)-H cytochrome c reductase in guinea pig polymorphonuclear leukocytes and alveolar macrophages during hyperoxia. J. clin. Invest. *58:* 1174 (1976).

19 Rüfer, R. und Stolz, C.: Inaktivierung von alveolären Oberflächenfilmen durch Erniedrigung der Oberflächenspannung der Hypophase. Pflügers Arch. ges. Physiol. *307:* 89 (1969).

20 Salin, M.L. and McCord, J.M.: Superoxide dismutases in polymorphonuclear leucozytes. J. clin. Invest. *54:* 1009 (1974).

21 Suttorp, N.; Seeger, W.; Wolf, H. und Neuhof, H.: Tierexperimentelles Modell einer Schocklunge durch standardisierte Mikroembolisation der Lungenstrombahn von wachen Kaninchen mit inerten Humanalbuminpartikeln (in preparation).

22 Stocks, J.; Gutteridge, J.M.C.; Sharp, R.J., and Dormandy, T.L.: The inhibition of lipid autoxidation by human serum and its relation to serum proteins and α-tocopherol. Clin. Sci. mol. Med. *47:* 223 (1974).

23   Tappel, A.L.: Lipid peroxidation damage to cell components. Fed. Proc. *32:* 1870
      (1973).
24   Thomas, H.v.; Müller, P.K., and Lyman, R.L.: Lipoperoxidation of lung lipids in rats
      exposed to nitrogen dioxide. Science *159:*532 (1968).
25   Weibel, E.R.: Toxische Auswirkungen erhöhter Sauerstoffspannung auf die Lunge; in
      Lungenveränderungen bei Langzeitbeatmung, p. 214 Wiemers und Scholler, (Thieme,
      Stuttgart 1973).
26   Weiss, S.J.; Rustagi, P.K.; and Lobuglio, A.F.: Human granulocyte generation of
      hydroxyl radical. J. exp. Med. *147:* 316 (1978).
27   Williams, R.A.; Rhoades, R.A., and Adams, W.S.: The response of lung tissue and
      surfactant to nitrogen dioxide exposure. Archs intern. Med. *128:*101 (1971).
28   Winsel, K.; Lachmann, B. und Reutgen, H.: Der Anti-Atelekase Faktor der Lunge. II.
      Mitteilung. Z. Erkr. Atm. *139:* 167 (1974).

Dr. med. Hellmut Wolf, Zentrum für Innere Medizin, Justus-Liebig-Universität,
Klinikstrasse 36, D-6300 Giessen (FRG)

Discussion

Answering a question by Dr. *Morley,* Dr. *Wolf* said that the premature babies were deficient in tocopherol. The doses he had used were very high, 100 times higher than the physiological dosis of tocopherol. Dr. *Zänker* came back to the fact that superoxide dismutase is one
of the fastest enzymes within the lung. Before showing a deficiency of this enzyme Dr. *Wolf*
cannot conclude that tocopherol will have any antioxidant effect in his experiments.
Dr. *Geiger* had used tocopherol in oxygen toxic rats and had not seen any effect compared
with the tocopherol rats in the control group.

Subject Index